SPRINGER PROCEEDINGS IN PHYSICS 118

SPRINGER PROCEEDINGS IN PHYSICS

106 **Modern Trends in Geomechanics**
Editors: W. Wu and H.S. Yu

107 **Microscopy of Semiconducting Materials**
Proceedings of the 14th Conference,
April 11–14, 2005, Oxford, UK
Editors: A.G. Cullis and J.L. Hutchison

108 **Hadron Collider Physics 2005**
Proceedings of the 1st Hadron
Collider Physics Symposium,
Les Diablerets, Switzerland,
July 4–9, 2005
Editors: M. Campanelli, A. Clark,
and X. Wu

109 **Progress in Turbulence II**
Proceedings of the iTi Conference
in Turbulence 2005
Editors: M. Oberlack, G. Khujadze,
S. Guenther, T. Weller, M. Frewer, J. Peinke,
S. Barth

110 **Nonequilibrium Carrier Dynamics in Semiconductors**
Proceedings
of the 14th International Conference,
July 25–29, 2005, Chicago, USA
Editors: M. Saraniti, U. Ravaioli

111 **Vibration Problems ICOVP 2005**
Editors: E. Inan, A. Kiris

112 **Experimental Unsaturated Soil Mechanics**
Editor: T. Schanz

113 **Theoretical and Numerical Unsaturated Soil Mechanics**
Editor: T. Schanz

114 **Advances in Medical Engineering**
Editor: T.M. Buzug

115 **X-Ray Lasers 2006**
Proceedings
of the 10th International Conference,
August 20–25, 2006, Berlin, Germany
Editors: P.V. Nickles, K.A. Janulewicz

116 **Lasers in the Conservation of Artworks**
LACONA VI Proceedings,
Vienna, Austria, Sept. 21–25, 2005
Editors: J. Nimmrichter, W. Kautek,
M. Schreiner

117 **Advances in Turbulence XI**
Proceedings of the 11th EUROMECH
European Turbulence Conference,
June 25–28, 2007, Porto, Portugal
Editors: J.M.L.M. Palma and A. Silva Lopes

118 **The Standard Model and Beyond**
Proceedings of the 2nd International
Summer School in High Energy Physics,
Muğla, 25–30 September 2006
Editors: T. Aliev, N.K. Pak, and M. Serin

Volumes 81–105 are listed at the end of the book.

T. Aliev N.K. Pak M. Serin
(Eds.)

The Standard Model and Beyond

Proceedings of the 2nd International Summer School in High Energy Physics, Muğla, 25–30 September 2006

With 50 Figures

Springer

Takhmasib Aliev
Namık Kemal Pak
Meltem Serin

MiddleEast Technical University (METU)
Physics Department
06531 Ankara, Turkey
E-mail: serinm@metu.edu.tr

ISSN 0930-8989

ISBN 978-3-642-09278-7 e-ISBN 978-3-540-73621-9

Springer is a part of Springer Science+Business Media.

springer.com

Softcover reprint of the hardcover 1st edition 2008

Cover: eStudio Calamar Steinen

Preface

This is the Proceedings of the Second International Summer School in High Energy Physics, focusing on "Standard Model and Beyond", we have held in Mḡla, Turkey, on 25–30 September 2006.

The participants were mostly graduate and post-graduate students (about 50) working on the experimental and phenomenological problems. The students were mostly Turkish, and there were some from Azerbaijan.

In addition to excellent tutorial lectures on the Standard Model and the Grand Unification, there were exposures on the current developments on heavy flavor physics, the physics at LHC, recent experimental data coming from BABAR and BELLE, as well as neutrino oscillations data, which were presented by the senior spoke-persons of TEXONO Collaboration.

Encouraged by the success of this year's event, as well as the preceding one, we are planning to make this school an annual event, with substantial international student participation, and with special focus to those from central Asian countries.

We think we owe a lot to the venue where the event has been held. The special location of the convention center where the event has been held, Gökova, within half an hour driving distance from Ephesus and Milethos, has been superbly praised by the international participants. These great heritages from antique Ionia are the home of some of our founding fathers of western civilization and science, from Thales to Heraklitos. There is another event which already gained the status of a world event, namely the Gökova School on Geometry and Topology which uses the same venue. This success story plays a weighted role in our encouragement. This very special spot of the world deserves to be paid special tribute when it comes to looking for a special place for intellectual gatherings in Europe.

Clearly it is not an easy matter to elevate such events into a world status, like Erice, Les Houches, Cargese, etc. Continuity and publicity are two most crucial factors among others. There is another factor, which is a prerequisite of the first factor, namely the commitment of the sponsors.

For domestic administrative publicity purpose, as well as for reference purposes for graduate and post-graduate students, we have decided to publish the proceedings in the book form. Another reason to publish the proceedings in the book form is to draw the attention of the international peer community, to this wonderful spot, one of the cradles of western civilizations. We are sure that this particular volume will help us to achieve the goals we are aiming at concerning the future faith of our conference.

The event was partially supported by funds from Turkish National Agencies, Turkish Atomic Energy Authority (TAEK) and The Scientific and Technical Research Council of Turkey (TÜBİTAK), which we thankfully acknowledge. This proceeding could not have been published without the financial support of TAEK. We are very grateful to Okay Çakıroğlu, the President of TAEK, for providing this support.

The Conference would not have been possible without the generous services provided by the Muḡla University. We thank the Administration of the University as well as many volunteers from the Physics Department that helped us to organize social events and the excursions to the antique cities of Ephesus and Miletos.

We thank B. Ögel, the vice president of Middle East Technical University, and the administration of the Physics Department for a wide spectrum of support, from printing services to providing conference stationary.

We are most grateful to our speakers for their excellent talks and friendly cooperation to prepare their manuscripts under a pressing schedule, and the Springer Verlag publishers for their very positive, and cooperative approach.

We thank the members of organizing committee for their valuable contributions in shaping the scientific program and various organizational services. We also thank the young researchers who interactively contributed to the running of discussion hours and tutorials. There are contributed talks which we could not unfortunately include in these proceeding due to some internal constraints; we acknowledge these sincerely. We further thank all participants for their interest and commitment to the school.

We thank last but not the least, to our students, Ç Özkan, S. Sekmen and H. Gamsızkan for their untiring assistances.

Ankara,
May 2007

T. Aliev, N.K. Pak,
M. Serin

Contents

Contents

List of Contributors

Buğra Borasoy
Helmholtz-Institut für Strahlen-und Kernphysik, Universität Bonn
Nussallee 14-16, D-53115 Bonn
Germany
borasoy@itkp.uni-bonn.de

Durmus A. Demir
Department of Physics, İzmir Institute of Technology, IZTECH
TR35430, İzmir, Turkey
demir@physics.iztech.edu.tr

George W.S. Hou
Department of Physics
National Taiwan University
Taipei, Taiwan 10617, R.O.C.
wshou@phys.ntu.edu.tw

Livio Lanceri
Dipartimento di Fisica
Università di Trieste
and INFN, Sezione di Trieste
livio.lanceri@ts.infn.it

Sh. Mamedov
Institute for Physical Problems
Baku State University
Z. Khalilov 23, Baku
AZ/1148, Azerbaijan
Sh_-Mamedov@bsu.az

Goran Senjanović
ICTP, Trieste, Italy
goran@ictp.it

Henry T. Wong
Institute of Physics
Academia Sinica
Taipei 11529, Taiwan
htwong@phys.sinica.edu.tw

Introduction to Chiral Perturbation Theory

Buḡra Borasoy

Helmholtz-Institut für Strahlen-und Kernphysik, Universität Bonn,
Nussallee 14-16, D-53115 Bonn, Germany borasoy@itkp.uni-bonn.de

A brief introduction to chiral perturbation theory, the effective field theory of quantum chromodynamics at low energies, is given.

1 Introduction

The strong interactions are described by quantum chromodynamics (QCD), a local non-abelian gauge theory. The QCD Lagrangian comprises quark and gluon fields which carry *color* charges and interact with coupling strength g. The renormalized coupling g depends on the momentum at which the measurement is performed and decreases as the momentum scale Q is increased. This behavior is referred to as *running* of the strong coupling constant $\alpha_s(Q) = g^2(Q)/(4\pi)$. The coupling α_s decreases for large momenta and the theory becomes *asymptotically free* with quasi-free quarks and gluons [1].

In this regime of QCD perturbation theory in α_s converges. For small momenta, on the other hand, α_s is large so that quarks and gluons arrange themselves in strongly bound clusters to form hadrons, e.g., protons, neutrons, pions, kaons, etc. In order to describe the physics of hadrons at low energies, perturbation theory is not useful because α_s is large. This is illustrated for $\pi\pi$ scattering in Fig. 1 where both sample diagrams—along with infinitely many other contributions—are equally important, although the right diagram appears at a much higher order in the perturbative series in α_s.

Alternative model-independent approaches are required in the non-perturbative regime of QCD. These are provided either by QCD lattice simulations which are a numerical solution to QCD or—at low energies—by chiral perturbation theory, the effective field theory of QCD. In the first case the QCD path integral in Euclidean space-time is evaluated numerically via Monte Carlo sampling, see e.g. [2]. In the latter case, one makes use of the fact that at low energies the relevant, effective degrees of freedom are hadrons rather than quarks and gluons which are not observed as free particles.

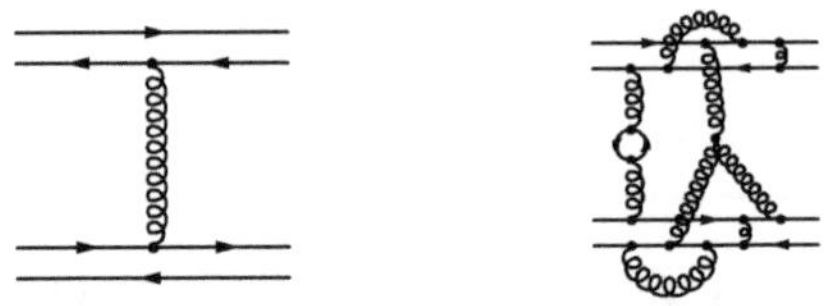

Fig. 1. Two sample diagrams which contribute to $\pi\pi$ scattering. Solid and curly lines denote quarks and gluons, respectively.

It is thus convenient to replace in the low-energy limit the QCD Lagrangian by an effective Lagrangian which is formulated in terms of the effective degrees of freedom, i.e. pions, kaons, eta, etc. The corresponding field theoretical formalism is called chiral perturbation theory (ChPT) [3–5].

In these lectures a brief introduction to ChPT is presented emphasizing some basic principles and a few simple applications. It is not intended to provide a detailed review of ChPT, in particular we restrict ourselves to the purely mesonic sector and do not consider baryons. For more comprehensive reviews the reader is referred to [6].

This paper is organized as follows. In the next section some well-known examples and basic principles of effective field theories in general are presented. Section 3 describes the construction principles for the chiral effective Lagrangian. Higher orders and loops are discussed in Sec. 4.

2 Effective Field Theories

The basic idea of an effective field theory is to treat the active, light particles as relevant degrees of freedom, while the heavy particles are frozen and reduced to static sources. The dynamics are described by an effective Lagrangian which is formulated in terms of the light particles and incorporates all important symmetries and symmetry-breaking patterns of the underlying fundamental theory.

2.1 Scattering of Light by Light in QED at Very Low Energies

The Lagrangian of quantum electrodynamics (QED) is given by

$$\mathcal{L}_{\text{QED}} = \mathcal{L}_0 + \mathcal{L}_{\text{int}} \tag{1}$$

with the free part

$$\mathcal{L}_0 = \bar{\psi}\,(i\partial\!\!\!/ - m)\,\psi - \frac{1}{4}F_{\mu\nu}F^{\mu\nu} \tag{2}$$

and the interaction piece

$$\mathcal{L}_{\text{int}} = -e\bar{\psi}A\!\!\!/\psi\ . \tag{3}$$

Fermion and photon fields are denoted by ψ and A_μ, respectively, $F_{\mu\nu} = \partial_\mu A_\nu - \partial_\nu A_\mu$ is the field strength tensor, and a gauge fixing term has been omitted for brevity.

Consider light by light scattering at very low photon energies $\omega \ll m$. In this instance, electrons (and positrons) cannot be produced in the final state, but contribute instead via virtual processes. The calculation of the lowest order diagram which is given by a single electron loop, Fig. 2, is straightforward but cumbersome. However, at very low energies the amplitude for light by light scattering is equally reproduced by the effective Lagrangian [7,8]

$$\mathcal{L}_{\text{eff}} = -\frac{1}{4}F_{\mu\nu}F^{\mu\nu} + \frac{e^4}{1440\,\pi^2 m^4}\left[(F_{\mu\nu}F^{\mu\nu})^2 + \frac{7}{16}(F_{\mu\nu}\tilde{F}^{\mu\nu})^2\right] + \ldots \qquad (4)$$

which only contains the field strength tensor $F_{\mu\nu}$ and its dual counterpart $\tilde{F}_{\mu\nu} = \epsilon_{\mu\nu\rho\sigma}F^{\rho\sigma}$ as explicit degress of freedom. The ellipsis denotes corrections to this Lagrangian involving more derivatives which arise from the energy expansion of the original one-loop diagram in powers of ω/m. Moreover, the coefficients of the operators in the effective Lagrangian receive corrections of higher orders in e^2 through multiloop diagrams.

It is instructive to illustrate the conversion to the effective field theory with Feynman diagrams. By treating at very low photon energies the electrons as heavy static sources the electron propagators of the electron loop in QED "shrink" to a single point. This gives rise to 4-photon contact interactions which correspond to the vertices of the effective Lagrangian, see Fig. 3.

A significant property is that the U(1) gauge symmetry of the underlying QED Lagrangian is maintained by the effective Lagrangian, Eq. (4), since the building blocks $F_{\mu\nu}F^{\mu\nu}$ and $F_{\mu\nu}\tilde{F}^{\mu\nu}$ are both gauge invariant. As we will see below, invariance under the relevant symmetries is an important constraint in constructing effective Lagrangians.

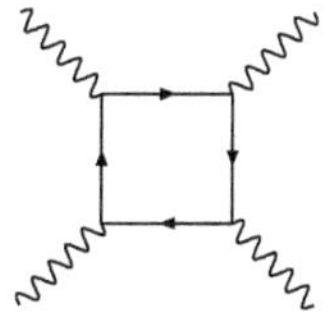

Fig. 2. Light by light scattering to lowest order. The wavy and solid lines denote the photons and electrons, respectively.

Fig. 3. The one-loop diagram of QED is replaced in the effective theory by 4-photon contact interactions.

Fig. 4. At low energies the single W boson exchange reduces to a four-quark contact interaction.

2.2 Weak Interactions at Very Low Energies

A second well-known example of an effective field theory is encountered in weak interactions. Consider the amplitude for the flavor changing weak process at lowest order from single W boson exchange

$$\mathcal{A} = \left(\frac{ig}{\sqrt{2}}\right)^2 V_{us}V_{ud}^* \left(\bar{u}\gamma^\mu \tfrac{1-\gamma_5}{2} s\right) \left(\bar{d}\gamma^\nu \tfrac{1-\gamma_5}{2} u\right) \left(\frac{-ig_{\mu\nu}}{p^2 - M_W^2}\right) , \tag{5}$$

where V_{ij} are elements of the Kobayashi-Maskawa mixing matrix and the W propagator is given in Feynman gauge. In the limit of small momentum transfer, $p^2 \ll M_W^2$, the W propagator can be expanded in p^2/M_W^2 such that the amplitude is approximated by the local interaction

$$\mathcal{A} = \frac{i}{M_W^2} \left(\frac{ig}{\sqrt{2}}\right)^2 V_{us}V_{ud}^* \left(\bar{u}\gamma^\mu \tfrac{1-\gamma_5}{2} s\right) \left(\bar{d}\gamma_\mu \tfrac{1-\gamma_5}{2} u\right) + \mathcal{O}\left(\frac{p^2}{M_W^4}\right) . \tag{6}$$

Diagrammatically this approximation is illustrated in Fig. 4, where the contact interaction arises from the effective Lagrangian

$$\mathcal{L}_{\text{eff}} = -2\sqrt{2} G_F V_{us}V_{ud}^* \left(\bar{u}\gamma^\mu \tfrac{1-\gamma_5}{2} s\right) \left(\bar{d}\gamma_\mu \tfrac{1-\gamma_5}{2} u\right) \tag{7}$$

with the Fermi constant $G_F = g^2/\left(4\sqrt{2}M_W^2\right)$.

2.3 Chiral Symmetry in QCD

As mentioned above, the relevant symmetries of the underlying theory must also be maintained by the effective field theory. In this section, we will study the (approximate) chiral symmetry of QCD. The QCD Lagrangian reads in compact notation

$$\mathcal{L}_{\text{QCD}} = \bar{q}\left(i\gamma^\mu D_\mu - m_q\right) q - \frac{1}{2}\text{Tr}_c\left(G^{\mu\nu}G_{\mu\nu}\right) , \tag{8}$$

where $q^T = (u, d, s, c, b, t)$ comprises the six quark flavors, $D_\mu = \partial_\mu - igG_\mu$ is the covariant derivative, G_μ the gluon fields, and $G_{\mu\nu} = \partial_\mu G_\nu - \partial_\nu G_\mu - ig\left[G_\mu, G_\nu\right]$ the gluon field strength tensor. Tr_c denotes the trace in color space. The Dirac field q is a 72-component object; each of the 6 quark flavors appears in 3 different colors and has 4 spinor components.

The quarks can be grouped into light and heavy flavors according to their masses: the u, d, s quarks are substantially lighter than the c, b, t quarks [9]. Hence, the limit of massless light quarks, $m_u = m_d = m_s = 0$, the so-called chiral limit, seems to be a reasonable approximation and can be improved by treating the light quark masses as perturbations. The c, b, t quarks, on the other hand, can be treated at low energies as infinitely heavy and the only active degrees of freedom are those associated with the light u, d, s quarks.

It is straightforward to see that in the chiral limit the QCD Lagrangian has an extra symmetry. In this limit, the relevant part of $\mathcal{L}_{\mathrm{QCD}}$ is (we use the same notation for simplicity)

$$\mathcal{L}_{\mathrm{QCD}} = \sum_{q=u,d,s} \bar{q} i \gamma_\mu D^\mu q - \frac{1}{2}\mathrm{Tr}_c(G_{\mu\nu}G^{\mu\nu}) \ . \tag{9}$$

Here, q represents a one-flavor quark field. By introducing right- and left-handed quark fields

$$q_{R/L} = \frac{1}{2}(1 \pm \gamma_5)\, q \tag{10}$$

one arrives at

$$\mathcal{L}_{\mathrm{QCD}} = \sum_{q=u,d,s} (\bar{q}_L i \gamma_\mu D^\mu q_L + \bar{q}_R i \gamma_\mu D^\mu q_R) - \frac{1}{2}\mathrm{Tr}_c(G_{\mu\nu}G^{\mu\nu}) \ . \tag{11}$$

Independent transformations of the right- and left-handed quark fields

$$q_R \to R\, q_R \ , \qquad\qquad q_L \to L\, q_L \tag{12}$$

with $R \in \mathrm{SU}(3)_R$, $L \in \mathrm{SU}(3)_L$ leave the massless QCD Lagrangian invariant. This invariance is referred to as $\mathrm{SU}(3)_L \times \mathrm{SU}(3)_R$ chiral symmetry of massless QCD. One observes that the gluon interactions do not change the helicity of quarks but the quark mass term does.

Due to Noether's theorem an immediate consequence of a continuous symmetry of a Lagrangian is the existence of a conserved current J_μ with $\partial_\mu J^\mu = 0$. The corresponding charge

$$Q(t) = \int d^3x\ J_0(t, \mathbf{x}) \tag{13}$$

is time-independent, i.e. $dQ/dt = 0$. Familiar examples are the invariance of the Lagrangian with regard to translations in time and space and rotations which imply, respectively, conservation of energy, momentum and angular momentum. At the operator level, the conserved charges commute with the Hamiltonian.

In the chiral limit of QCD the conserved currents of chiral symmetry are

$$L^a_\mu = \sum_{q=u,d,s} \bar{q}_L \gamma_\mu \frac{\lambda^a}{2} q_L \ , \qquad\qquad R^a_\mu = \sum_{q=u,d,s} \bar{q}_R \gamma_\mu \frac{\lambda^a}{2} q_R \tag{14}$$

with the Gell-Mann matrices λ^a. The invariant charges Q_L^a, Q_R^a generate the algebra of $\mathrm{SU}(3)_L$ and $\mathrm{SU}(3)_R$, respectively. It is useful to define the combinations

$$Q_V^a = Q_R^a + Q_L^a \; ; \qquad Q_A^a = Q_R^a - Q_L^a \tag{15}$$

which have a different behavior under parity

$$Q_V^a \to Q_V^a \; ; \qquad Q_A^a \to -Q_A^a \; . \tag{16}$$

Consider an eigenstate $|\psi\rangle$ of H_{QCD} (in the chiral limit)

$$H_{\mathrm{QCD}}|\psi\rangle = E|\psi\rangle \; . \tag{17}$$

The states $Q_V^a|\psi\rangle$ and $Q_A^a|\psi\rangle$ have the same energy E but opposite parity. Thus for each positive parity state there should be a negative parity state with equal mass. This pattern is, however, not observed in the particle spectrum [9]. For example, the light pseudoscalar ($J^P = 0^-$) mesons, (π, K, η), have a considerably lower mass than the scalar ($J^P = 0^+$) mesons.

The solution to this paradoxon is provided by the Nambu-Goldstone realization of chiral symmetry [10] which asserts that the QCD vacuum, $|0\rangle$, is not invariant under the action of the axial charges

$$Q_V^a|0\rangle = 0 \qquad Q_A^a|0\rangle \neq 0 \; . \tag{18}$$

The chiral $SU(3)_L \times SU(3)_R$ symmetry of the QCD Hamiltonian is said to be *spontaneously* broken down to $SU(3)_V$. Spontaneous breakdown of a symmetry takes place if the full symmetry group of the Hamiltonian is not shared by the vacuum.

Another example of spontaneous symmetry breakdown occurs in ferromagnets. For temperatures above the Curie temperature, $T > T_c$, the magnetic dipoles are randomly oriented. As soon as the temperature falls below the Curie temperature T_c spontaneous magnetization occurs and the dipoles are aligned in some arbitrary direction. Spontaneous symmetry breakdown takes also place for the $\mathrm{SU}(2)_L \times \mathrm{U}(1)$ symmetry of the electroweak interactions.

In general, spontaneous breakdown of a continuous symmetry has important consequences. Goldstone's theorem states that a spontaneously broken continuous symmetry implies massless spinless particles: the Goldstone bosons. In the case of massless QCD, the eight axial charges Q_A^a create states $|\phi\rangle = Q_A|0\rangle$ which are energetically degenerate with the vacuum $|0\rangle$ since

$$H|\phi\rangle = HQ_A|0\rangle = Q_A H|0\rangle = 0 \; . \tag{19}$$

This gives rise to eight massless pseudoscalar mesons. The axial charges Q_A^a acting on any particle state generate Goldstone bosons, e.g. an energy eigenstate $|\psi\rangle$ is degenerate with the multi-particle state $Q_A^a|\psi\rangle$ which resolves the paradoxon from above.

The eight lightest hadrons are indeed the pseudoscalars $\pi^\pm, \pi^0, K^\pm, K^0, \bar{K}^0$, η with masses $m_\pi \approx 138$ MeV, $m_K \approx 495$ MeV and $m_\eta \approx 547$ MeV [9]. Since the nonzero masses of the light quarks break chiral symmetry explicitly the Goldstone bosons are not exactly massless. However, the explicit breaking can be considered to be small and treated perturbatively. In the limit of vanishing quark masses, $m_u, m_d, m_s \to 0$, the Goldstone boson masses approach zero, $m_\pi, m_K, m_\eta \to 0$, while all other hadrons remain massive in the chiral limit and are separated from the ground state roughly by a characteristic gap

$$\Delta \sim M_{\text{proton}} \sim 1\,\text{GeV} \,. \tag{20}$$

In the remainder of this section, it is demonstrated that only the spontaneous breakdown of a *continuous* symmetry gives rise to Goldstone bosons, whereas in the case of a discrete symmetry Goldstone bosons are not generated.

Discrete symmetry case

Consider the Lagrangian density with a scalar field ϕ

$$\mathcal{L} = \frac{1}{2}\partial_\mu\phi\partial^\mu\phi + \frac{1}{2}m^2\phi^2 - \frac{1}{4}\lambda\phi^4 \,. \tag{21}$$

The Lagrangian is invariant under the discrete symmetry of reflections, $\phi \to -\phi$. The corresponding potential is given by

$$V(\phi^2) = -\frac{1}{2}m^2\phi^2 + \frac{1}{4}\lambda\phi^4 \,, \tag{22}$$

and since the energy must be bound from below the coupling λ is positive. The coefficient m^2, on the other hand, is not constrained. There are two possible cases depending on the sign of m^2 as illustrated in Fig. 5. For $m^2 < 0$ there is a unique minimum at $\phi = 0$, but for $m^2 > 0$ the potential $V(\phi^2)$ is minimized by two possible ground state fields $\phi = \pm\sqrt{m^2/\lambda}$. In the quantum field theoretical language this implies that the field ϕ develops a vacuum expectation value

$$\langle 0|\phi|0\rangle = \pm\sqrt{\frac{m^2}{\lambda}} \,. \tag{23}$$

Hence, there are two possible vacua but each vacuum is not invariant under reflection symmetry, i.e. the theory is spontaneously broken. Massless Goldstone bosons, however, do not appear.

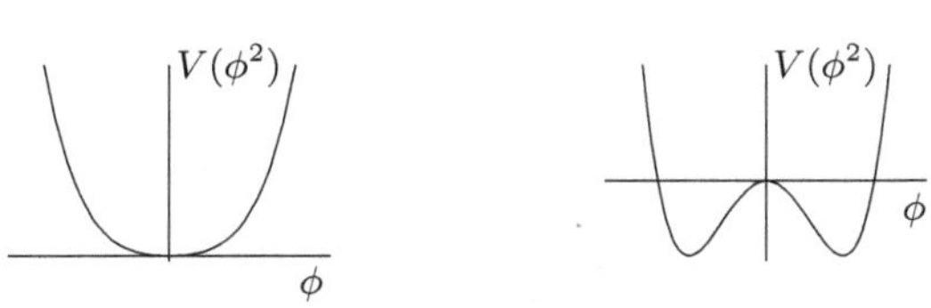

Fig. 5. Potential $V(\phi^2)$, Eq. (22), for $m^2 < 0$ (left) and $m^2 > 0$ (right).

Continuous symmetry case

Consider now the Lagrangian with two scalar fields σ and π

$$\mathcal{L} = \frac{1}{2}(\partial_\mu \sigma)^2 + \frac{1}{2}(\partial_\mu \pi)^2 - V(\sigma^2 + \pi^2) \tag{24}$$

with V defined as in Eq. (22). It exhibits an O(2) symmetry; continuous transformations of the type

$$\begin{pmatrix} \sigma \\ \pi \end{pmatrix} \to \begin{pmatrix} \cos\alpha & \sin\alpha \\ -\sin\alpha & \cos\alpha \end{pmatrix} \begin{pmatrix} \sigma \\ \pi \end{pmatrix} \tag{25}$$

leave the Lagrangian invariant. The extrema of the corresponding potential V are determined by the equations

$$\begin{aligned} \frac{dV}{d\sigma} &= \sigma[-m^2 + \lambda(\sigma^2 + \pi^2)] = 0 \ , \\ \frac{dV}{d\pi} &= \pi[-m^2 + \lambda(\sigma^2 + \pi^2)] = 0 \ . \end{aligned} \tag{26}$$

For $m^2 > 0$ the minima are at $\sigma^2 + \pi^2 = m^2/\lambda$ and related to each other through O(2) rotations. Any point on the circle of minima may be chosen to be the true vacuum $|0\rangle$. One may take, e.g.,

$$\langle 0|\sigma|0\rangle = \sqrt{\frac{m^2}{\lambda}} \ ; \qquad \langle 0|\pi|0\rangle = 0 \ . \tag{27}$$

Clearly, the O(2) symmetry of the Lagrangian is spontaneously broken by the vacuum state. Small oscillations around this vacuum state can be described by shifting the σ field

$$\sigma' \equiv \sigma - \sqrt{\frac{m^2}{\lambda}} \tag{28}$$

so that the Lagrangian reads in terms of the new fields (up to an irrelevant constant)

$$\begin{aligned} \mathcal{L} = & \frac{1}{2}\left(\partial_\mu \sigma'\right)^2 + \frac{1}{2}\left(\partial_\mu \pi\right)^2 - m^2 \sigma'^2 \\ & -\lambda \sqrt{\frac{m^2}{\lambda}} \sigma' \left(\sigma'^2 + \pi^2\right) - \frac{1}{4}\lambda \left(\sigma'^2 + \pi^2\right)^2 \ . \end{aligned} \tag{29}$$

With this choice of coordinates the mass term for the π field has disappeared and the π becomes massless. The π field is then interpreted as a polar angle oscillation around the vacuum which does not cost any energy. A Goldstone boson has been created through spontaneous breakdown of the continuous O(2) symmetry in the original Lagrangian.

3 Construction of the Chiral Effective Lagrangian

In this section we outline the construction principles for the chiral effective Lagrangian. The chiral SU(3) Lagrangian is in general a function of the Goldstone boson (GB) fields ($\pi^0, \pi^\pm, K^\pm, K^0, \bar{K}^0, \eta$). In order to construct the effective Lagrangian, we must first know the interaction between the GBs.

To this aim, we recall from the previous section that the eight axial charges Q_A^a do not annihilate the vacuum, $Q_A^a|0\rangle \neq 0$. The states $Q_A^a|0\rangle \neq 0$ are associated with the GBs $\phi^a = (\pi, K, \eta)$. This implies non-vanishing matrix elements of the axial vector current A_μ^a

$$\langle 0|A_\mu^a(x)|\phi^b(p)\rangle = ie^{-ip\cdot x}\, p_\mu\, \delta^{ab}\, f_a \tag{30}$$

(no summation over a). The decay constant f_a measures the strength with which the Goldstone boson ϕ^a decays via the axial vector current A_μ^a into the hadronic vacuum. The decay constants are extracted experimentally from weak decays of the GBs, e.g., $\pi^+ \to l^+\nu_l$ yields $f_\pi = 92.4$ MeV [11].

Taking the divergence of Eq. (30) leads to

$$\langle 0|\partial^\mu A_\mu^a(0)|\phi^b(p)\rangle = \delta^{ab} m_a^2 f_a \ . \tag{31}$$

In the chiral limit, the axial vector current is conserved, $\partial^\mu A_\mu^a = 0$, so that $m_a^2 = 0$ as required by Goldstone's theorem. In the real world, however, chiral $SU(3)_L \times SU(3)_R$ symmetry is explicitly broken by the finite quark masses m_u, m_d, m_s and the axial vector current is not conserved. One introduces the GB field operators Φ^a with the normalization $\langle 0|\Phi^a(0)|\phi^b(p)\rangle = \delta^{ab}$. Eq. (31) can then be rewritten as

$$\langle 0|\partial^\mu A_\mu^a(0)|\phi^b(p)\rangle = m_a^2 f_a \langle 0|\Phi^a(0)|\phi^b(p)\rangle \ . \tag{32}$$

At the operator level, this is the hypothesis of the partially conserved axial vector current (PCAC)

$$\partial^\mu A_\mu^a = m_a^2 f_a \Phi^a \ . \tag{33}$$

The axial currents can thus be employed as interpolating fields for the Goldstone bosons and identity (33) implies a vanishing interaction between the GBs at zero momentum. Consider to this end, e.g., the matrix element (suppressing flavor indices)

$$\mathcal{M}_\mu(p_1, p_2, p_3) = \langle \phi(p_2)\phi(p_3)|A_\mu(0)|\phi(p_1)\rangle \ . \tag{34}$$

The amplitude $\mathcal{M}_\mu$ contains two parts: contributions with no GB poles and contributions where the axial current generates a GB pole, see Fig. 6. In the chiral limit, the matrix element has the decomposition

$$\mathcal{M}_\mu(p_1, p_2, p_3) = \frac{f q_\mu}{q^2} T(p_1, p_2, p_3, q) + R_\mu \ , \tag{35}$$

Fig. 6. Contributions to $\mathcal{M}_\mu$ in Eq. (34) with (left) and without (right) a GB pole. The wavy and solid lines denote the axial vector current and the Goldstone bosons, respectively.

where $q = -p_1 - p_2 - p_3$, T is the GB-GB scattering matrix element, f the decay constant in the chiral limit, and R_μ is non-singular as $q_\mu \to 0$ by definition. Contracting both sides with q^μ yields

$$0 = q^\mu \mathcal{M}_\mu(p_1, p_2, p_3) = fT(p_1, p_2, p_3, q) + q^\mu R_\mu \ . \tag{36}$$

In the limit $q_\mu \to 0$ one obtains

$$T(p_1, p_2, p_3, q) = 0 \ . \tag{37}$$

The GBs do not interact at vanishing momenta.

At low but finite energies, the interaction between GBs can be expanded in powers of small momenta. Consider for example the GB-GB scattering matrix T which can be written as a function of the three Mandelstam variables $s = (p_1 + p_2)^2$, $t = (p_1 + p_3)^2$ and $u = (p_1 + q)^2$. Its low energy expansion reads

$$T(s, t, u) = f_1 s + g_1 t + h_1 u + \dots \tag{38}$$

with momentum-independent expansion coefficients f_i, g_i, h_i. The chiral effective Lagrangian is also ordered according to the low energy expansion. Powers of GB momenta in the amplitude correspond to powers of derivatives on GB fields in the Lagrangian. The ordering of the effective Lagrangian in increasing powers of derivatives is called *chiral ordering* or *chiral power counting*.

Next, we would like to investigate how GB fields are represented in the chiral Lagrangian. To this aim, we shall study the transformation properties of the GBs under chiral transformations.

Let G be the group of chiral SU(3)$_L\times$ SU(3)$_R$ transformations. For a given representation of G the GB fields transform according to

$$\phi \to \phi' = F(g, \phi) \ , \qquad g \in G \tag{39}$$

with the representation property

$$F(g_1, F(g_2, \phi)) = F(g_1 g_2, \phi) \ . \tag{40}$$

Consider group elements $h \in G$ which leave the "origin", i.e. the vacuum, invariant, $F(h, 0) = 0$. Obviously, these elements form a subgroup H: for

$h_1, h_2 \in H$ it follows that $h_1 h_2 \in H$. H is equivalent to the subgroup $\mathrm{SU}(3)_V$ which leaves the vacuum invariant.

The function

$$g \to F(g,0) = F(gh,0) \qquad h \in H \tag{41}$$

maps the coset space G/H onto the space of GB fields. This mapping is invertible since $F(g_1,0) = F(g_2,0)$ implies $g_1^{-1} g_2 \in H$. As the dimension of the coset space is equal to the number of Goldstone boson fields, the GBs can be identified with elements of G/H. The Goldstone boson fields are said to *live* on the coset space $\mathrm{SU}(3)_L \times \mathrm{SU}(3)_R/\mathrm{SU}(3)_V$.

Any $g \in G$ can be decomposed as $g = qh$ with $q \in G/H$ and $h \in H$. The choice of representatives in the coset space G/H is arbitrary. Possible choices are for example

$$g = (g_L, g_R) = (1, g_R g_L^{-1})(g_L, g_L) \equiv qh \tag{42}$$

or

$$g = (g_L, g_R) = (g_L g_R^{-1}, 1)(g_R, g_R) \equiv q'h' \ . \tag{43}$$

If we pick, e.g., the latter choice then the action of G on G/H is given by

$$(L,R)(g_L g_R^{-1}, 1) = (L g_L g_R^{-1}, R) = (L g_L g_R^{-1} R^{-1}, 1)(R,R) \ . \tag{44}$$

The Goldstone bosons are then summarized by the matrix-valued field $U = g_L g_R^{-1}$ which transforms under chiral transformations as

$$U(x) \to U'(x) = L U(x) R^{-1} = L U(x) R^\dagger \tag{45}$$

for $L/R \in \mathrm{SU}(3)_{L/R}$. The exponential representation is convenient for $U \in \mathrm{SU}(3)$

$$U = \exp\left(\frac{i}{f}\phi^a \lambda^a\right) \ , \tag{46}$$

where λ^a are the generators of SU(3)

$$\phi = \phi^a \lambda^a = \sqrt{2} \begin{pmatrix} \frac{1}{\sqrt{2}}\pi^0 + \frac{1}{\sqrt{6}}\eta & \pi^+ & K^+ \\ \pi^- & -\frac{1}{\sqrt{2}}\pi^0 + \frac{1}{\sqrt{6}}\eta & K^0 \\ K^- & \bar{K}^0 & -\frac{2}{\sqrt{6}}\eta \end{pmatrix} . \tag{47}$$

The chiral effective Lagrangian for QCD is written in terms of the GB fields which are collected in the matrix-valued field U

$$\mathcal{L}_{\text{eff}} = \mathcal{L}_{\text{eff}}(U, \partial U, \partial^2 U, \ldots) \ . \tag{48}$$

The effective Lagrangian shares the same symmetries with QCD: C, P, T, Lorentz invariance and, in particular, chiral $\mathrm{SU}(3)_L \times \mathrm{SU}(3)_R$ symmetry. As outlined above, the chiral Lagrangian is expanded in chiral powers which are related (in the chiral limit) to the number of derivatives acting on the GB fields. The chiral power counting of the Lagrangian reads

$$\mathcal{L}_{\text{eff}} = \mathcal{L}_{\text{eff}}^{(0)} + \mathcal{L}_{\text{eff}}^{(2)} + \mathcal{L}_{\text{eff}}^{(4)} + \dots \ . \tag{49}$$

Only even chiral powers arise since the Lagrangian is a Lorentz scalar which implies that tensor indices of derivatives appear in pairs. At each chiral order the effective Lagrangian must be invariant under chiral $\mathrm{SU}(3)_L \times \mathrm{SU}(3)_R$ transformations. At zeroth chiral order this invariance implies that $\mathcal{L}_{\text{eff}}^{(0)}$ can only be a function of $UU^\dagger = 1$. This amounts to an irrelevant constant in the Lagrangian which can be dropped.

At second order, the chiral invariant terms with two derivatives are

$$\mathcal{L}_{\text{eff}}^{(2)} = c_1 \langle \partial_\mu U^\dagger \partial^\mu U \rangle + c_2 \langle U^\dagger \Box U \rangle \ , \tag{50}$$

where $\langle \dots \rangle$ is the trace in flavor space. The second term can be reduced to the first one by partial integration; only one term remains at second chiral order

$$\mathcal{L}_{\text{eff}}^{(2)} = c_1 \langle \partial_\mu U^\dagger \partial^\mu U \rangle \ . \tag{51}$$

Since terms of zeroth chiral order have been dropped, the second chiral order is effectively the leading order (LO). We note the appearance of a coupling constant c_1, a so-called low-energy constant (LEC). It is fixed by expanding the matrix-valued field U in the GB fields ϕ

$$U = \exp\left(\frac{i}{f}\phi\right) = 1 + \frac{i}{f}\phi - \frac{1}{2f^2}\phi^2 + \mathcal{O}(\phi^3) \tag{52}$$

and requiring the standard kinetic term

$$\mathcal{L}_{\text{eff}}^{(2)} = \frac{1}{2}\partial_\mu \phi^a \partial^\mu \phi^a + \mathcal{O}(\phi^4) \tag{53}$$

which yields $c_1 = f^2/4$.

Therefore, the effective Lagrangian at LO reads

$$\mathcal{L}_{\text{eff}}^{(2)} = \frac{f^2}{4} \langle \partial_\mu U^\dagger \partial^\mu U \rangle \ . \tag{54}$$

At leading chiral order there is only one LEC (in the chiral limit) and chiral symmetry constrains all vertices with increasing number of GB fields in the LO Lagrangian.

The interpretation of the LEC f can be directly inferred by considering the Noether axial current of chiral symmetry for $\mathcal{L}_{\text{eff}}^{(2)}$

$$A_\mu^a = i\frac{f^2}{4} \langle \lambda^a \{\partial_\mu U, U^\dagger\} \rangle \ . \tag{55}$$

Upon comparison with the PCAC hypothesis

$$\langle 0|A_\mu^a(0)|\phi^b(p)\rangle = ip_\mu \delta^{ab} f_a \tag{56}$$

one confirms that f is the GB decay constant in the chiral limit.

As a first application we are now in a position to predict, e.g., $\pi\pi$ scattering at leading chiral order. The scattering amplitude has the decomposition

$$\begin{aligned} &\mathcal{M}(\pi^a(p_a)\,\pi^b(p_b) \to \pi^c(p_c)\,\pi^d(p_d)) \\ &\quad = \delta^{ab}\delta^{cd}A(s,t,u) + \delta^{ac}\delta^{bd}A(t,s,u) + \delta^{ad}\delta^{bc}A(u,t,s) \ , \end{aligned} \tag{57}$$

where a, b, c, d are flavor indices and $s = (p_a + p_b)^2$, $t = (p_a - p_c)^2$, $u = (p_a - p_d)^2$. Employing $\mathcal{L}_{\text{eff}}^{(2)}$ one calculates

$$A(s,t,u) = \frac{s}{f^2} \ . \tag{58}$$

Up to now, we have worked in chiral limit $m_u, m_d, m_s = 0$ where chiral symmetry is exact. In the real world the quark masses do not vanish and introduce an explicit breaking of chiral symmetry in $\mathcal{L}_{\text{QCD}}$

$$\mathcal{L}_{\text{QCD}} = \mathcal{L}_{\text{QCD}}^0 - \bar{q}\mathcal{M}q = \mathcal{L}_{\text{QCD}}^0 - \bar{q}_R\mathcal{M}q_L - \bar{q}_L\mathcal{M}q_R \ , \tag{59}$$

where $\mathcal{L}_{\text{QCD}}^0$ is the massless QCD Lagrangian and $\mathcal{M} = \text{diag}(m_u, m_d, m_s)$ the light quark mass matrix. The chiral symmetry breaking patterns induced by the light quark masses must be reproduced at the level of the effective field theory. To this end, we interpret the quark mass matrix as an external scalar source s

$$\bar{q}\mathcal{M}q = \bar{q}_L\mathcal{M}q_R + \bar{q}_R\mathcal{M}^\dagger q_L \to \bar{q}_L s q_R + \bar{q}_R s^\dagger q_L \ . \tag{60}$$

The external scalar source s is required to transform under chiral rotations as

$$s \to L s R^\dagger \ . \tag{61}$$

Obviously, this leaves the QCD Lagrangian invariant under chiral rotations and implies that the effective Lagrangian must also remain invariant in the presence of s. Hence, the chiral invariant effective Lagrangian is extended with s as an additional building block

$$\mathcal{L}_{\text{eff}}(U, \partial U, \partial^2 U, \ldots) \to \mathcal{L}_{\text{eff}}(U, \partial U, \partial^2 U, \ldots; s) \ . \tag{62}$$

Once the effective Lagrangian is constructed one can go back to the real world by setting $s = \mathcal{M}$. In (standard) chiral perturbation theory[1] the chiral counting rule is $s = \mathcal{M} = \mathcal{O}(p^2)$, i.e. the quark masses are booked as second

---

[1] We do not consider here the framework of generalized ChPT wherein $\mathcal{M} = \mathcal{O}(p)$ [12].

chiral order. In addition to the derivative expansion the effective Lagrangian is expanded in powers of s. At leading chiral order one obtains

$$\begin{aligned}\mathcal{L}_{\text{eff}}^{(2)} &= \frac{f^2}{4}\langle\partial_\mu U^\dagger \partial^\mu U\rangle + \frac{f^2}{2}B\langle sU^\dagger + Us^\dagger\rangle \\ &= \frac{f^2}{4}\langle\partial_\mu U^\dagger \partial^\mu U\rangle + \frac{f^2}{2}B\langle\mathcal{M}(U+U^\dagger)\rangle \ . \end{aligned} \tag{63}$$

We have introduced an additional constant B related to explicit chiral symmetry breaking.

The expansion of the symmetry-breaking part in powers of GB fields reads

$$\frac{f^2}{2}B\langle\mathcal{M}(U+U^\dagger)\rangle = (m_u+m_d+m_s)Bf^2 - \frac{1}{2}B\langle\mathcal{M}\phi^2\rangle + \mathcal{O}(\phi^4) \ . \tag{64}$$

The first term in this expansion is related to the vacuum expectation values of the scalar quark densities

$$\langle 0|\bar{q}q|0\rangle = \langle 0|\frac{\partial H_{\text{QCD}}}{\partial m_q}|0\rangle = -\langle 0|\frac{\partial \mathcal{L}_{\text{eff}}}{\partial m_q}|0\rangle = -f^2B + \mathcal{O}(m_q) \tag{65}$$

and B is the order parameter of spontaneous chiral symmetry breaking. At lowest order the quark condensates are degenerate

$$\langle 0|\bar{u}u|0\rangle = \langle 0|\bar{d}d|0\rangle = \langle 0|\bar{s}s|0\rangle = -f^2B \ . \tag{66}$$

The second term in the expansion in Eq. (64) is the mass term for the GB fields

$$\begin{aligned} m_{\pi^+}^2 &= 2B\hat{m} \\ m_{\pi^0}^2 &= 2B\hat{m} + \mathcal{O}\left[\frac{(m_u-m_d)^2}{m_s-\hat{m}}\right] \\ m_{K^+}^2 &= B(m_u+m_s) \\ m_{K^0}^2 &= B(m_d+m_s) \\ m_\eta^2 &= \frac{2}{3}B(\hat{m}+2m_s) + \mathcal{O}\left[\frac{(m_u-m_d)^2}{m_s-\hat{m}}\right] \end{aligned} \tag{67}$$

with $\hat{m} = \frac{1}{2}(m_u+m_d)$. The terms proportional to $(m_u-m_d)^2$ are due to π^0-η mixing. They constitute tiny corrections and are usually neglected. The mass relations in Eq. (67) explain why one counts $s = \mathcal{O}(p^2)$ (taking B to be of zeroth chiral order). For finite light quark masses we see that the GBs are not massless any longer (sometimes they are referred to as *pseudo*-GBs).

From the previous equalities one obtains immediately the Gell-Mann–Oakes–Renner relations

$$\begin{aligned} f_\pi^2 m_\pi^2 &= -2\hat{m}\langle 0|\bar{q}q|0\rangle + \mathcal{O}(m_q^2) \\ f_K^2 m_K^2 &= -(\hat{m}+m_s)\langle 0|\bar{q}q|0\rangle + \mathcal{O}(m_q^2) \\ f_\eta^2 m_\eta^2 &= -\tfrac{2}{3}(\hat{m}+2m_s)\langle 0|\bar{q}q|0\rangle + \mathcal{O}(m_q^2) \end{aligned} \tag{68}$$

and the Gell-Mann–Okubo mass relation

$$3m_\eta^2 = 4m_K^2 - m_\pi^2 \ . \tag{69}$$

To leading chiral order, the strong interactions at low energies are characterized by the two scales f and B with

$$\begin{aligned} f \simeq f_\pi &\simeq 93\,\text{MeV} \\ B &\simeq 1800\,\text{MeV} \ , \end{aligned} \tag{70}$$

where the value of B has been extracted from the sum-rule value $\langle 0|\bar{q}q|0\rangle = -(250\,\text{MeV})^3$ [13].

One important goal of chiral perturbation theory is to extract the light quark mass ratios from phenomenology. Note that the absolute values of the quark masses cannot be extracted since they depend on the QCD renormalization scale, but the QCD scale dependence cancels out in quark mass ratios. For example, from the leading order mass relations one obtains

$$\frac{m_u}{m_d} = \frac{(m_{K^+}^2 - m_{K^0}^2)_{\text{QCD}} + m_{\pi^0}^2}{(m_{K^0}^2 - m_{K^+}^2)_{\text{QCD}} + m_{\pi^0}^2} \ , \tag{71}$$

where the tiny contribution from π^0-η mixing and electromagnetic effects in the kaon mass difference have been neglected. Next, we employ Dashen's theorem [14] which asserts that electromagnetic contributions to the mass splittings $m_{K^+}^2 - m_{K^0}^2$ and $m_{\pi^+}^2 - m_{\pi^0}^2$ are identical in the chiral limit

$$\begin{aligned} (m_{K^+}^2 - m_{K^0}^2)_{\text{QCD}} &= (m_{K^+}^2 - m_{K^0}^2)_{\text{phys}} - (m_{K^+}^2 - m_{K^0}^2)_{\text{em}} \\ &= (m_{K^+}^2 - m_{K^0}^2)_{\text{phys}} - (m_{\pi^+}^2 - m_{\pi^0}^2)_{\text{em}} \\ &= (m_{K^+}^2 - m_{K^0}^2)_{\text{phys}} - (m_{\pi^+}^2 - m_{\pi^0}^2)_{\text{phys}} \ . \end{aligned} \tag{72}$$

The last equation is valid since the pion mass difference is (almost) entirely due to electromagnetic contributions. Insertion of this identity in the quark mass ratio yields [15]

$$\frac{m_u}{m_d} = \frac{m_{K^+}^2 - m_{K^0}^2 + 2m_{\pi^0}^2 - m_{\pi^+}^2}{m_{K^0}^2 - m_{K^+}^2 + m_{\pi^+}^2} = 0.55 \ . \tag{73}$$

At leading order, the possibility $m_u = 0$ which would lead to a solution of the strong CP problem is excluded. In a similar manner one finds [15]

$$\frac{m_s}{m_d} = \frac{m_{K^0}^2 + m_{K^+}^2 - m_{\pi^0}^2}{m_{K^0}^2 - m_{K^+}^2 + m_{\pi^+}^2} = 20.1 \ . \tag{74}$$

These leading order ratios are, of course, subject to higher order corrections.

We note in passing that due to the inclusion of finite quark masses the $\pi\pi$ scattering amplitude, Eq. (58), is modified according to

$$A(s,t,u) = \frac{s - m_\pi^2}{f^2} = \frac{s - m_\pi^2}{f_\pi^2} + \mathcal{O}(p^4) \ . \tag{75}$$

In the remainder of this section we extend the effective field theoretical formalism to include external vector and axial-vector fields. This is necessary for the calculation of Green functions of quark currents, e.g., for the electromagnetic form factor of the pion

$$\langle \pi(p')|J_\mu^{\rm em}|\pi(p)\rangle = \langle \pi(p')|\bar{q}\gamma_\mu Q q|\pi(p)\rangle \ , \tag{76}$$

where $Q = \frac{1}{3}{\rm diag}(2,-1,-1)$ is the quark charge matrix. Green functions of quark currents are conveniently calculated by applying the external field method. (As a matter of fact, we have already encountered this technique for the scalar quark densities $\langle 0|\bar{q}q|0\rangle$ which are obtained by differentiating with respect to the external field s.)

To this aim, one extends $\mathcal{L}_{\rm QCD}$ in the presence of external fields

$$\mathcal{L}_{\rm QCD} = \mathcal{L}^0_{\rm QCD} + \bar{q}\gamma_\mu(v^\mu + \gamma_5 a^\mu)q - \bar{q}(s + i\gamma_5 p)q \ , \tag{77}$$

where v^μ, a^μ, s, and p are hermitian matrices denoting the external vector, axial-vector, scalar, and pseudoscalar fields, respectively. The coupling of external photon fields $\mathcal{A}_\mu$ to the quarks, e.g., is given by the prescription $v_\mu = -eQ\mathcal{A}_\mu$. Green functions of electromagnetic currents are then obtained by functional derivatives of the generating functional $Z[v,a,s,p]$ with respect to $\mathcal{A}_\mu$. Couplings to weak currents are also collected in v_μ and a_μ.

Due to the inclusion of electromagnetic and weak currents global chiral symmetry is promoted to a local one. The Lagrangian $\mathcal{L}_{\rm QCD}$ in Eq. (77) remains invariant under the local chiral $\mathrm{SU}(3)_L\times\mathrm{SU}(3)_R$ transformations

$$\begin{aligned} q_R &\rightarrow Rq_R \\ q_L &\rightarrow Lq_L \\ r_\mu = v_\mu + a_\mu &\rightarrow Rr_\mu R^\dagger + iR\partial_\mu R^\dagger \\ l_\mu = v_\mu - a_\mu &\rightarrow Ll_\mu L^\dagger + iL\partial_\mu L^\dagger \\ s + ip &\rightarrow L(s+ip)R^\dagger \ . \end{aligned} \tag{78}$$

Local chiral symmetry must also be maintained by the extended effective Lagrangian in the presence of external fields

$$\mathcal{L}_{\rm eff}(U,\partial U,\partial^2 U,\ldots) \rightarrow \mathcal{L}_{\rm eff}(U,\partial U,\partial^2 U,\ldots;v,a,s,p) \ . \tag{79}$$

The local nature of chiral symmetry requires the introduction of a covariant derivative

$$\partial_\mu U \to D_\mu U = \partial_\mu U - i l_\mu U + i U r_\mu \tag{80}$$

which transforms under chiral rotations as

$$D_\mu U \to L(D_\mu U)R^\dagger \ . \tag{81}$$

In addition, the vector and axial-vector fields v_μ, a_μ enter the effective Lagrangian also through the non-Abelian field strength tensors, $R_{\mu\nu}, L_{\mu\nu}$,

$$\begin{aligned} R_{\mu\nu} &= \partial_\mu r_\nu - \partial_\nu r_\mu - i[r_\mu, r_\nu] \\ L_{\mu\nu} &= \partial_\mu l_\nu - \partial_\nu l_\mu - i[l_\mu, l_\nu] \end{aligned} \tag{82}$$

with the transformation properties

$$R_{\mu\nu} \to R\, R_{\mu\nu}\, R^\dagger \ , \qquad L_{\mu\nu} \to L\, L_{\mu\nu}\, L^\dagger \ . \tag{83}$$

4 Higher Orders and Loops

In the previous section we have outlined the construction principles for the chiral effective Lagrangian. The effective Lagrangian is ordered in powers of Goldstone boson momenta and masses, a chiral power counting scheme is established. The leading order effective Lagrangian $\mathcal{L}_{\text{eff}}^{(2)}$ was constructed explicitly and sample calculations have been performed at the leading tree level order. In this section we focus on the inclusion of next-to-leading order (NLO) contributions. On the one hand, these are the tree level diagrams with vertices from $\mathcal{L}_{\text{eff}}^{(4)}$. On the other hand, loop diagrams with vertices from $\mathcal{L}_{\text{eff}}^{(2)}$ must be taken into account.

The chiral expansion of the effective Lagrangian reads

$$\mathcal{L}_{\text{eff}} = \mathcal{L}_{\text{eff}}^{(2)} + \mathcal{L}_{\text{eff}}^{(4)} + \dots \ . \tag{84}$$

The most general effective Lagrangian at fourth chiral order, $\mathcal{L}_{\text{eff}}^{(4)}$, which respects the relevant symmetries of QCD is [5][2]

$$\begin{aligned} \mathcal{L}_{\text{eff}}^{(4)} &= L_1 \langle D_\mu U^\dagger D^\mu U\rangle^2 + L_2 \langle D_\mu U^\dagger D_\nu U\rangle \langle D^\mu U^\dagger D^\nu U\rangle \\ &\quad + L_3 \langle D_\mu U^\dagger D^\mu U D_\nu U^\dagger D^\nu U\rangle + L_4 \langle D_\mu U^\dagger D^\mu U\rangle \langle \chi U^\dagger + U\chi^\dagger\rangle \\ &\quad + L_5 \langle D_\mu U^\dagger D^\mu U(\chi^\dagger U + U^\dagger \chi)\rangle + L_6 \langle \chi U^\dagger + U\chi^\dagger\rangle^2 \\ &\quad + L_7 \langle \chi U^\dagger - U\chi^\dagger\rangle^2 + L_8 \langle \chi U^\dagger \chi U^\dagger + U\chi^\dagger U\chi^\dagger\rangle \\ &\quad - i L_9 \langle R^{\mu\nu} D_\mu U^\dagger D_\nu U + L^{\mu\nu} D_\mu U D_\nu U^\dagger\rangle + L_{10} \langle R_{\mu\nu} U^\dagger L^{\mu\nu} U\rangle \end{aligned} \tag{85}$$

with $\chi \equiv 2B(s + ip)$. Ten low-energy constants enter with unknown values which must be determined from phenomenology. At next-to-next-to-leading

[2] We have omitted two terms that are not accessible from experiment and only serve the purpose of absorbing loop divergences.

order even 90 unknown LECs enter [16]. Due to the proliferation of unknown couplings in the effective Lagrangian ChPT seems to lose its predictive power. However, the situation is ameliorated since only a few LECs contribute to a particular process. There is no need to determine all LECs of the effective Lagrangian for a given process. Moreover, the contributions from higher chiral orders are suppressed as they appear with additional powers of small momenta or Goldstone boson masses. Therefore in practice only the LECs from lower orders are required.

At fourth chiral order, contributions from $\mathcal{L}_{\text{eff}}^{(4)}$ to the pseudoscalar meson masses stem solely from tree diagrams, see Fig. 7. Corrections from $\mathcal{L}_{\text{eff}}^{(4)}$ to $\pi\pi$ scattering at $\mathcal{O}(p^4)$ also contribute through tree diagrams, Fig. 8.

As already indicated above, tree diagrams are not the whole story. At next-to-leading order loop diagrams must be included as well. The loops also restore unitarity of the S-matrix which can be seen as follows. Unitarity of the S-matrix implies

$$SS^\dagger = 1 \; . \tag{86}$$

Inserting the usual decomposition $S = 1 + iT$ yields

$$-i(T - T^\dagger) = T^\dagger T \; . \tag{87}$$

Since T is symmetric due to time reversal invariance the last relation simplifies to

$$2\,\mathrm{Im}\,T = T^\dagger T \; . \tag{88}$$

Tree diagrams do not have an imaginary piece and thus cannot satisfy Eq. (88), i.e., unitarity is violated if only tree diagrams are taken into account. However, unitarity of the S-matrix is perturbatively restored by including loop diagrams which have imaginary parts.

As a simple example of a loop calculation, we sketch the evaluation of the one-loop contributions to the GB self-energies. To this end, we expand the U fields in $\mathcal{L}_{\text{eff}}^{(2)}$

Fig. 7. Tree diagram contribution to the self-energy of the Goldstone bosons. The square denotes a vertex from $\mathcal{L}_{\text{eff}}^{(4)}$.

Fig. 8. Tree diagram contribution to $\pi\pi$ scattering at order order $\mathcal{O}(p^4)$. The square denotes a vertex from $\mathcal{L}_{\text{eff}}^{(4)}$.

Fig. 9. Tadpole contribution to the self-energy of the Goldstone bosons. The vertex stems from $\mathcal{L}_{\text{eff}}^{(2)}$.

$$
\begin{aligned}
\mathcal{L}_{\text{eff}}^{(2)} &= \frac{f^2}{4}\langle D_\mu U^\dagger D^\mu U\rangle + \frac{f^2}{2}B\langle \mathcal{M}(U^\dagger + U)\rangle \\
&= \mathcal{L}_{\text{kin}} + \mathcal{L}_{\text{int}} \\
&= \frac{1}{4}\langle \partial_\mu \phi \partial^\mu \phi\rangle - \frac{1}{2}B\langle \mathcal{M}\phi^2\rangle \\
&\quad + \frac{1}{48 f^2}\langle [\partial_\mu \phi, \phi][\partial^\mu \phi, \phi]\rangle + \frac{1}{24 f^2}B\langle \mathcal{M}\phi^4\rangle + \dots
\end{aligned} \tag{89}
$$

Interaction vertices with four GB fields arise which yield contributions to the meson masses via tadpoles, see Fig. 9.

The calculation of the tadpole diagram involves the loop integration

$$
\int \frac{d^4 k}{(2\pi)^4} \frac{i}{k^2 - m_\phi^2 + i\epsilon}, \tag{90}
$$

where m_ϕ is the GB mass in the loop. This loop integral is quadratically ultraviolet divergent and a regularization scheme is necessary which maintains the symmetries of the theory, in particular chiral symmetry. One must be careful when employing cutoff regularization schemes since these introduce an additional massive cutoff scale which may spoil chiral symmetry [17]. In this respect, a convenient mass-independent regularization scheme is provided by dimensional regularization. The basic idea of dimensional regularization is as follows. The dimension of the loop integration is changed to an arbitrary dimension $d \in \mathbb{R}$. For d small enough the integral remains ultraviolet finite and can be performed explicitly. Afterwards, the analytic continuation back to $d = 4$ dimensions is performed.

To be more explicit, consider the tadpole integral in d dimensions

$$
I = \mu^{4-d} \int \frac{d^d k}{(2\pi)^d} \frac{i}{k^2 - m_\phi^2 + i\epsilon}. \tag{91}
$$

The regularization scale μ has been introduced to maintain the mass dimension of I for arbitrary space-time dimension d. The calculation of I is performed in a straightforward manner

$$
\begin{aligned}
I &= \mu^{4-d} \int \frac{d^d k}{(2\pi)^d} \frac{i}{k^2 - m_\phi^2 + i\epsilon} = \mu^{4-d} \int \frac{d^d k_{\text{E}}}{(2\pi)^d} \frac{1}{k_{\text{E}}^2 + m_\phi^2} \\
&= \mu^{4-d} \int \frac{d^d k_{\text{E}}}{(2\pi)^d} \int_0^\infty d\lambda \; \exp\left\{-\lambda\left(k_{\text{E}}^2 + m_\phi^2\right)\right\}
\end{aligned}
$$

$$= \frac{\mu^{4-d}}{(2\pi)^d} \int_0^\infty d\lambda \ \exp\left(-\lambda m_\phi^2\right) \int d^d k_{\mathrm{E}} \ \exp\left(-\lambda k_{\mathrm{E}}^2\right)$$

$$= \frac{\mu^{4-d}}{(2\pi)^d} \int_0^\infty d\lambda \ \exp\left(-\lambda m_\phi^2\right) \left(\frac{\pi}{\lambda}\right)^{d/2}$$

$$= (4\pi)^{-\frac{d}{2}} m_\phi^2 \left(\frac{m_\phi}{\mu}\right)^{d-4} \Gamma(1-\frac{d}{2}) \ . \tag{92}$$

A Wick rotation to Euclidean space has been performed in the first line. Expanding the result in $d-4$ yields

$$I = m_\phi^2 \left\{ 2L + \frac{1}{16\pi^2} \ln\left(\frac{m_\phi^2}{\mu^2}\right) \right\} + \mathcal{O}(d-4) \tag{93}$$

with the divergent piece

$$L = \frac{\mu^{d-4}}{16\pi^2} \left[\frac{1}{d-4} - \frac{1}{2} \left(\ln 4\pi + 1 + \Gamma'(1)\right) \right] \ . \tag{94}$$

The tadpole integral I is decomposed into two parts: a non-analytic term from the GB infrared singularities proportional to the "chiral log" $\ln(m_\phi^2/\mu^2)$ and a piece L which diverges for $d \to 4$.

A complete one-loop calculation for GB self-energies with vertices from $\mathcal{L}_{\text{eff}}^{(2)}$ reveals that divergent parts proportional to L are of fourth chiral order and analytic in the quark masses and GB momenta. Hence, the divergent components can be compensated by renormalizing the LECs of fourth chiral order

$$L_i = L_i^r + \lambda_i L \tag{95}$$

such that the sum of loops and LECs (including fourth chiral order) remains finite for $d \to 4$ and can be evaluated in four space-time dimensions.

The above considerations for one-loop contributions to the self-energy can be generalized to other processes and higher loops. Consider a connected L-loop diagram for a given physical process. The chiral order (or chiral dimension) D of a loop integral counts dimensions of GB masses and momenta (utilizing a mass-independent regularization scheme such as, e.g., dimensional regularization) and is given by

$$D = 2 + 2L + \sum_k (k-2) N_k \ , \tag{96}$$

where L is the number of loops and N_k the number of vertices of order $\mathcal{O}(p^k)$.

Vertices from the lowest order Lagrangian $\mathcal{L}_{\text{eff}}^{(2)}$ do not alter the chiral dimension, while vertices from higher orders increase D. Each loop also increases the chiral dimension, i.e. the loopwise expansion corresponds with the

Fig. 10. Leading $\mathcal{O}(p^2)$ contribution to $\pi\pi$ scattering. The vertex stems from $\mathcal{L}_{\text{eff}}^{(2)}$.

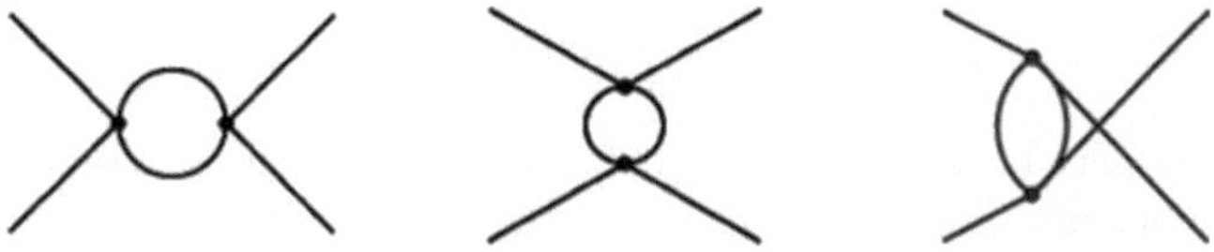

Fig. 11. One-loop diagrams for $\pi\pi$ scattering with vertices from $\mathcal{L}_{\text{eff}}^{(2)}$.

chiral expansion. From Eq. (96) one infers that at leading chiral order only tree diagrams with vertices of $\mathcal{L}_{\text{eff}}^{(2)}$ contribute. At next-to-leading order both tree diagrams with exactly one fourth order vertex and one-loop diagrams with vertices of second chiral order contribute. Two-loop diagrams start contributing at sixth chiral order and so forth.

As an explicit example of the chiral expansion we consider the Feynman diagrams for $\pi\pi$ scattering up to one-loop order. At lowest order, $\mathcal{O}(p^2)$, a tree diagram with a lowest order vertex contributes, see Fig. 10. At next-to-leading order, $\mathcal{O}(p^4)$, both tree diagrams with vertices from $\mathcal{L}_{\text{eff}}^{(4)}$ (Fig. 8) and one-loop diagrams with lowest order vertices (Fig. 11) contribute. The corresponding loop integrals are of the type

$$\int \frac{d^d l}{(2\pi)^d} \frac{p_1 \cdot p_2 \;\; p_3 \cdot p_4}{[l^2 - m_a^2 + i\epsilon] \; [(l + p_1 + p_2)^2 - m_b^2 + i\epsilon]} \; . \tag{97}$$

The construction of the most general chiral effective Lagrangian compliant with the symmetries and usage of a regularization scheme which respects these symmetries guarantees that appearing divergences can be absorbed in the LECs. Therefore, quantum effects do not violate chiral symmetry. For a mass-independent regularization scheme the occuring divergences will be of higher chiral order than the involved vertices. In this way, a systematic renormalization procedure emerges. The complete renormalization program at one- and two-loop order has been carried out by employing a background field method and heat kernel techniques [4, 5, 18].

The chiral expansion is ordered in increasing powers of GB masses and momenta. Each additional loop produces a factor of $1/f^2$ from the expansion of $U = \exp(i\phi/f)$ multiplied by the loop factor $1/(4\pi)^2$. Since each loop increases the chiral dimensionality D by two, inverse powers of the scale $4\pi f \simeq 4\pi f_\pi \simeq 1.2\,\text{GeV}$ enter in the chiral expansion. The scale of spontaneous chiral symmetry breakdown is thus

$$\Lambda_\chi = 4\pi f_\pi \simeq 1.2\,\text{GeV} \; . \tag{98}$$

Moreover, variation of the regularization scale μ allows for shuffling finite contributions from loops to LECs and vice versa. This suggests that contributions from renormalized LECs have roughly the same order of magnitude and, thus, the same expansion scale. Based on this naive estimate, the natural expansion parameter of the chiral amplitudes is expected to be

$$\frac{m_\eta^2}{16\pi^2 f_\pi^2} \simeq 0.22 \ . \tag{99}$$

However, in practice substantial higher order corrections which are not in accord with this rule may occur in the chiral expansion.

Let us focus now on the values of the renormalized constants which are not fixed by chiral symmetry. They must be calculated either directly from QCD or fixed from experimental data. In the first case one tries to extract the values of the LECs from lattice QCD simulations, but only a smaller set of LECs has been extracted so far using this procedure, see e.g. [19]. The second option, i.e. the direct comparison of the chiral amplitudes with available experimental data, has been applied more extensively. The phenomenological values and sources for the renormalized coupling constants L_i^r at the regularization scale $\mu = M_\rho$ are presented in Table 1. Once the LECs are determined one can predict further observables at the same chiral order.

Consider, e.g., the extraction of L_5. The LEC L_5 governs SU(3) breaking in the GB decay constants

$$\frac{f_K}{f_\pi} = 1 + \frac{4L_5^r}{f^2}(m_K^2 - m_\pi^2) + \text{chiral logs} \ . \tag{100}$$

Comparison with the experimental value [9]

$$\left(\frac{f_K}{f_\pi}\right)_{\text{exp}} = 1.22 \tag{101}$$

fixes L_5^r.

Table 1. Phenomenological values and sources for the renormalized coupling constants $L_i^r(\mu = M_\rho)$ [20].

i	$L_i^r(\mu = M_\rho) \times 10^3$	source
1	0.4 ± 0.3	$\pi\pi \to \pi\pi, K_{e4}$
2	1.35 ± 0.3	$\pi\pi \to \pi\pi, K_{e4}$
3	-3.5 ± 1.1	$\pi\pi \to \pi\pi, K_{e4}$
4	-0.3 ± 0.5	Zweig rule
5	1.4 ± 0.5	f_K/f_π
6	-0.2 ± 0.3	Zweig rule
7	-0.4 ± 0.2	Gell-Mann–Okubo
8	0.9 ± 0.3	$m_{K^0}^2 - m_{K^+}^2$
9	6.9 ± 0.7	$\langle r^2 \rangle_V^\pi$
10	-5.5 ± 0.7	$\pi \to e\nu\gamma$

The values of the L_i can be understood in terms of meson resonance exchange [21]. For resonance masses considerably larger than the involved momenta, $M_R^2 \gg p^2$, the resonance propagators shrink to a single point and produce contact interactions of the chiral Lagrangian

$$\frac{1}{p^2 - M_R^2} \to -\frac{1}{M_R^2} + \mathcal{O}(p^2) . \tag{102}$$

Diagrammatically this is depicted in Fig. 12. Resonance saturation is based on the assumption that the $L_i^r(\mu = M_\rho)$ are practically saturated by resonance exchange (chiral duality). This principle appears to work very well, see Table 2. One also observes that whenever spin-1 resonances contribute they dominate the resonance exchange (chiral vector meson dominance).

After setting up the NLO Lagrangian we are now in a position to calculate the light quark mass ratios at next-to-leading order. The $\mathcal{O}(p^4)$ expressions for the light quark mass ratios are

$$\frac{m_K^2}{m_\pi^2} = \frac{m_s + \hat{m}}{m_u + m_d} \left[1 + \Delta_M + \mathcal{O}(m_q^2)\right] \tag{103}$$

and

$$\frac{(m_{K^0}^2 - m_{K^+}^2)_{\mathrm{QCD}}}{m_K^2 - m_\pi^2} = \frac{m_d - m_u}{m_s - \hat{m}} \left[1 + \Delta_M + \mathcal{O}(m_q^2)\right] \tag{104}$$

Fig. 12. In the static limit of heavy resonance masses (represented by the double line) vertices of the chiral effective Lagrangian are produced.

Table 2. Phenomenological values $L_i^r(\mu = M_\rho)$ in comparison with the values L_i^{res} from resonance saturation [21].

i	$L_i^r(\mu = M_\rho) \times 10^3$	$L_i^{\mathrm{res}} \times 10^3$
1	0.4 ± 0.3	0.6
2	1.35 ± 0.3	1.2
3	-3.5 ± 1.1	-3.0
4	-0.3 ± 0.5	0.0
5	1.4 ± 0.5	1.4
6	-0.2 ± 0.3	0.0
7	-0.4 ± 0.2	-0.3
8	0.9 ± 0.3	0.9
9	6.9 ± 0.7	6.9
10	-5.5 ± 0.7	-6.0

with the same chiral correction

$$\Delta_M = \frac{8}{f^2}[2L_8^r - L_5^r]\ (m_K^2 - m_\pi^2) + \text{chiral logs}\ . \tag{105}$$

Combining both results one obtains the parameter-free relation

$$Q^2 \equiv \frac{m_s^2 - \hat{m}^2}{m_d^2 - m_u^2} = \frac{m_K^2}{m_\pi^2} \frac{m_K^2 - m_\pi^2}{m_{K^0}^2 - m_{K^+}^2 + m_{\pi^+}^2 - m_{\pi^0}^2}\ , \tag{106}$$

where Dashen's theorem [14] has been employed

$$(m_{K^0}^2 - m_{K^+}^2)_{\text{QCD}} = m_{K^0}^2 - m_{K^+}^2 + m_{\pi^+}^2 - m_{\pi^0}^2\ . \tag{107}$$

This yields

$$Q_{\text{Dashen}} \simeq 24.1\ . \tag{108}$$

It is argued in the literature that chiral corrections to Dashen's theorem may decrease Q by up to 10% [22]. In fact, dispersive calculations of the decay $\eta \to 3\pi$ also suggest a slightly lower value [23], but the exact value for Q is still under lively discussion.

We close this section with two remarks.

SU(2) chiral perturbation theory

In this presentation, we have assumed that the strange quark mass m_s is light and in the chiral regime. The appropriate framework then is SU(3) ChPT. Since the group SU(2) is a subgroup of SU(3), the SU(3) Lagrangian is also valid for chiral SU(2) relevant for the pions ($\pi^\pm, \pi^0$). If, on the other hand, m_s is considered to be large, kaons and the eta are heavy particles and may be integrated out such that only pionic degrees of freedom remain [5]. The result is a pure SU(2) chiral Lagrangian with less LECs [4].

Axial U(1) anomaly

Finally, we would like to comment on the axial U(1) anomaly of the strong interactions. At the classical level, the QCD Lagrangian exhibits a $\mathrm{U}(3)_L \times \mathrm{U}(3)_R$ symmetry

$$\mathrm{U}(3)_L \times \mathrm{U}(3)_R = \mathrm{SU}(3)_L \times \mathrm{SU}(3)_R \times \mathrm{U}(1)_V \times \mathrm{U}(1)_A \tag{109}$$

which comprises the $\mathrm{SU}(3)_L \times \mathrm{SU}(3)_R$ symmetry considered so far. Under the transformations

$$q_R \to R q_R, \quad q_L \to L q_L, \qquad R, L \in \mathrm{U}(3) \tag{110}$$

the classical QCD Lagrangian remains invariant. The $\mathrm{U}(1)_V$ symmetry corresponds with baryon number conservation and is usually neglected. In contrast, the $\mathrm{U}(1)_A$ symmetry of the classical Lagrangian is broken at the quantum level. Under axial U(1) transformations the path integral picks up an additional contribution from the fermion determinant such that the full quantum theory does not exhibit the axial U(1) symmetry [24]. This is the so-called axial U(1) anomaly.

If the axial U(1) symmetry were not broken by the anomaly, it would imply the existence of another isoscalar 0^- Goldstone boson with a small mass [25]

$$m \leq \sqrt{3} m_\pi \, . \tag{111}$$

The lightest pseudoscalar meson and singlet under $\mathrm{SU}(3)_V$ is the η' with a mass

$$m_{\eta'} = 958\,\mathrm{MeV} \sim \Lambda_\chi \gg m_{\pi,K,\eta} \tag{112}$$

close to the scale of spontaneous chiral symmetry breakdown, Λ_χ. The η' is not a Goldstone boson and remains massive in the chiral limit. Hence, it is not included explicitly in conventional ChPT although its effects are hidden in coupling constants of the Lagrangian [5].

However, in the limit $N_c \to \infty$, where N_c is the number of colors, the axial U(1) anomaly vanishes and a nonet of Goldstone bosons is generated, (π, K, η, η'), which now includes the η' as ninth Goldstone boson with a mass comparable to the other GBs. Stated differently, the extra mass of the η' is induced by the axial anomaly. Starting from the large N_c limit it is possible to include the η' systematically in the effective theory [5, 26]. The chiral amplitudes can be expanded simultaneously both in chiral powers and powers of $1/N_c$ which establishes a rigorous counting scheme [27]. Besides, by adding the η' in the effective Lagrangian strong CP violation is automatically included and allows for the calculation of strong CP violating effects, such as the electric dipole moment of the neutron [28].

Acknowledgements

It is a pleasure to thank the organizers of the summer school for creating a very pleasant and stimulating atmosphere in a beautiful environment. I also thank Peter Bruns and Robin Nißler for a careful reading of the manuscript.

References

1. D. J. Gross and F. Wilczek: Phys. Rev. Lett. **30**, 1343 (1973);
   H. D. Politzer: Phys. Rev. Lett. **30**, 1346 (1973)
2. I. Montvay and G. Münster, “Quantum fields on a lattice,” Cambridge University Press, Cambridge (1994);
   R. Gupta: [arXiv:hep-lat/9807028]

3. S. Weinberg: Physica **96** A, 327 (1979)
4. J. Gasser and H. Leutwyler: Ann. Phys. **158**, 142 (1984)
5. J. Gasser and H. Leutwyler: Nucl. Phys. B **250**, 465 (1985)
6. See e.g:
   V. Bernard, N. Kaiser and U.-G. Meißner: Int. J. Mod. Phys. E **4**, 193 (1995);
   V. Bernard and U.-G. Meißner: [arXiv:hep-ph/0611231];
   G. Ecker: Prog. Part. Nucl. Phys. **35**, 1 (1995);
   G. Ecker: [arXiv:hep-ph/0011026];
   J. Gasser: Lect. Notes Phys. **629**, 1 (2004);
   H. Leutwyler: [arXiv:hep-ph/9406283];
   H. Leutwyler: [arXiv:hep-ph/0008124];
   A. V. Manohar: [arXiv: hep-ph/9606222];
   U.-G. Meißner: Rept. Prog. Phys. **56**, 903 (1993);
   A. Pich: [arXiv:hep-ph/9806303];
   S. Scherer: Adv. Nucl. Phys. **27**, 277 (2003);
   S. Scherer and M. R. Schindler: [arXiv:hep-ph/0505265]
7. W. Heisenberg and H. Euler: Z. Physik **98**, 714 (1936)
8. J. Schwinger: Phys. Rev. **82**, 664 (1951)
9. W.-M. Yao *et al.* [Particle Data Group Collaboration]: J. Phys. G **33**, 1 (2006)
10. Y. Nambu: Phys. Rev. Lett. **4**, 380 (1960); Phys. Rev. **117**, 648 (1960);
    Y. Nambu and G. Jona-Lasinio: Phys. Rev. **122**, 345 (1961)
11. B. R. Holstein: Phys. Lett. B **244**, 83 (1990)
12. N. H. Fuchs, H. Sazdjian and J. Stern: Phys. Lett. B **269**, 183 (1991);
    M. Knecht, H. Sazdjian, J. Stern and N. H. Fuchs: Phys. Lett. B **313**, 229 (1993)
13. M. A. Shifman, A. I. Vainshtein and V. I. Zakharov: Nucl. Phys. B **147**, 385 (1979)
14. R. Dashen: Phys. Rev. **183**, 1245 (1969)
15. S. Weinberg: Trans. NY Acad. Sci. **38**, 185 (1977)
16. J. Bijnens, G. Colangelo and G. Ecker: JHEP **9902**, 020 (1999)
17. I. Gerstein, R. Jackiw, B. W. Lee and S. Weinberg: Phys. Rev. D **3**, 2486 (1971)
18. J. Bijnens, G. Colangelo and G. Ecker: Annals Phys. **280**, 100 (2000)
19. See, e.g., C. Aubin *et al.* [MILC Collaboration]: Phys. Rev. D **70**, 114501 (2004)
20. J. Bijnens, G. Ecker and J. Gasser: [arXiv:hep-ph/9411232]
21. G. Ecker, J. Gasser, A. Pich and E. de Rafael: Nucl. Phys. B **321**, 311 (1989);
    J. F. Donoghue, C. Ramirez and G. Valencia: Phys. Rev. D **39**, 1947 (1989)
22. J. F. Donoghue, B. R. Holstein and D. Wyler: Phys. Rev. D **47**, 2089 (1993);
    R. Baur and R. Urech: Phys. Rev. D **53**, 6552 (1996);
    J. Bijnens and J. Prades: Nucl. Phys. B **490**, 239 (1997)
23. J. Kambor, C. Wiesendanger and D. Wyler: Nucl. Phys. B **465**, 215 (1996);
    A. V. Anisovich and H. Leutwyler: Phys. Lett. B **375**, 335 (1996);
    B. V. Martemyanov and V. S. Sopov: Phys. Rev. D **71**, 017501 (2005)
24. W. A. Bardeen: Phys. Rev. **184**, 1848 (1969);
    K. Fujikawa: Phys. Rev. Lett. **42**, 1195 (1979)
25. S. Weinberg: Phys. Rev. D **11**, 3583 (1975)
26. E. Witten: Ann. Phys. (N.Y.) **128**, 363 (1980);
    P. di Vecchia and G. Veneziano: Nucl. Phys. B **171**, 253 (1980)
27. R. Kaiser and H. Leutwyler: Eur. Phys. J. C **17**, 623 (2000)
28. A. Pich and E. de Rafael: Nucl. Phys. B **367**, 313 (1991);
    B. Borasoy: Phys. Rev. D **61**, 114017 (2000)

Introduction to Higgs Sector: Theory and Phenomenology

Durmus A. Demir

Department of Physics, İzmir Institute of Technology, IZTECH,
TR35430, İzmir, Turkey demir@physics.iztech.edu.tr

1 Introduction

This review is based on lectures delivered at International Summer School on High Energy Physics, in Akyaka, Muğla, Turkey, September 2006. The goal of the review is to introduce junior graduate students, having a some knowledge of field theory, to the problem of mass generation and its physical consequences. The review starts with the statements and a toy model of mass generation in Sect. 2. In Sect. 3 a model of weak interactions is constructed and its predictions are compared with experimental results obtained so far. Given in Sect. 4 is search strategies for Higgs boson, the only remnant of the mechanism that gives mass to matter, by referring to model predictions and experimental facilities. Finally, in Sect. 5, given are a summary of results plus reasons for and expectations from models which rehabilitate the minimal model constructed in Sect. 3[1].

2 Mass Generation

The inertial mass of a particle is a measure of its resistance to external forces acting upon, and of its energy content when at rest. In field theory language, mass of a particle is given by the pole of its propagator – the probability amplitude for particle to travel with a given energy and momentum. What is nontrivial about mass is that it is correlated with the symmetry properties of the field. In other words, symmetries and mass of a field are two intimately related structures. For a clearer view of this connection, it may be instructive to examine the electromagnetic field. This field is strictly massless. In fact,

[1] For the ease of grasping the fundamental concepts, and learning about alternative viewpoints present in the literature the references are chosen mostly from books and review papers and/or e-prints.

it has to be so since a finite mass explicitly breaks gauge invariance of electromagnetism, and hence, violates conservation of electric charge. Therefore, one arrives at a definite conclusion about the mass of photon on the basis of electric charge conservation alone [1, 2].

The connection between symmetry and mass of a field is not special to electromagnetism. A massless fermion field $\psi(x)$ exhibits chiral invariance, that is, its left– and right–handed components

$$\psi_L \equiv \frac{1-\gamma_5}{2}\psi\,,\quad \psi_R \equiv \frac{1+\gamma_5}{2}\psi \tag{1}$$

do not mix, and its Lagrangian

$$\mathcal{L}_{fermion} = \overline{\psi} i \not{D} \psi \tag{2}$$

is invariant under

$$\psi(x) \rightarrow e^{i\alpha_c(x)\gamma_5}\,\psi(x) \tag{3}$$

with the conserved current $J_\mu = \overline{\psi_R}\gamma_\mu\psi_R - \overline{\psi_L}\gamma_\mu\psi_L$. A finite mass for fermion explicitly breaks chiral invariance and spoils conservation of current: $\partial^\mu J_\mu = 2im_\psi\overline{\psi}\gamma_5\psi \neq 0$ [1, 2].

The weak interactions, which govern several phenomena ranging from radioactivity to burning of sun, proceed with the exchange of massive vector bosons. The nature of this interaction can be revealed by examining the decay

$$n \rightarrow p + \left(W^- \rightarrow\right) e^- + \nu_e \tag{4}$$

which rests on the existence of a charged intermediate vector boson $W^{\pm}_\mu$. This vector particle must be massive for decay amplitude to reduce Fermi's empirical contact interaction

$$\frac{G_F}{\sqrt{2}}\,\overline{p_L}\,\gamma^\mu n_L\,\overline{\nu_e}\,\gamma_\mu e_L \tag{5}$$

at low energies. The correspondence requires $G_F/\sqrt{2} = g_W^2/(8M_W^2)$ where g_W is strength of $W^{\pm}$ coupling to charged currents, and M_W is its mass [1, 2].

This brief discussion of weak interactions should convince oneself that massive vector bosons do exist in Nature and they must be incorporated in any theory of fundamental interactions. This is not straightforward, however, since, as mentioned in the case of electromagnetism, a hard mass for a vector boson explicitly breaks the gauge invariance. For a clearer view of this point it may be instructive to examine a generic Abelian vector field $A_\mu(x)$ (gauge field of some U(1) invariance [1, 2]) which transforms as

$$A_\mu \rightarrow A_\mu - \frac{1}{g}\partial_\mu\alpha \tag{6}$$

where g is gauge coupling and $\alpha(x)$ is an arbitrary scalar field. If $A_\mu(x)$ is given a mass M_A then its Lagrangian

$$\mathcal{L} = -\frac{1}{4}F^{\mu\nu}F_{\mu\nu} + \frac{1}{2}M_A^2 A^\mu A_\mu\,, \tag{7}$$

with $F_{\mu\nu} \equiv \partial_\mu A_\nu - \partial_\nu A_\mu$ being the field strength tensor, transforms as

$$\mathcal{L} \rightarrow \mathcal{L} - \frac{1}{2g}M_A^2\left(2A^\mu\partial_\mu\alpha - \frac{1}{g}\partial^\mu\alpha\partial_\mu\alpha\right) \tag{8}$$

under (6). The excess terms, those proportional to M_A^2, in this expression show that vector boson mass explicitly breaks invariance under $U(1)$ symmetry.

This rather brief exercise shows how nontrivial it is to induce a mass for a vector boson. To have a clue of correct description it proves useful to focus on vector boson self energy

$$\Pi_{\mu\nu}\left(q^2\right) = \left(g_{\mu\nu} - \frac{q_\mu q_\nu}{q^2}\right)\Pi\left(q^2\right) \tag{9}$$

whose proportionality to the transverse projector $q^2 g_{\mu\nu} - q_\mu q_\nu$ follows from Ward identity $q^\mu\Pi_{\mu\nu}\left(q^2\right) = 0$ meaning current conservation. The scalar self energy $\Pi\left(q^2\right)$ is generated by interactions of A_μ with itself and other fields. One obvious contribution is provided by the vector boson mass term in (7):

$$\Pi_{\mu\nu}^{mass}\left(q^2\right) = iM_A^2 g_{\mu\nu} \tag{10}$$

It is clear that to generate part of the projector proportional to $q_\mu q_\nu/q^2$ we obviously need a strictly massless scalar field (to identify $1/q^2$ factor with its propagator) which couples to A_μ via its derivative (to generate $q_\mu q_\nu$ in the numerator). This observation is guided by interaction between $\alpha(x)$ and $A_\mu(x)$ given in (8). Naming this massless scalar as ϕ, letting it couple to A_μ via $M_A^2 A^\mu\partial_\mu\phi$, and interpreting M_A^2 at the vertex to be a background field one obtains

$$\Pi_{\mu\nu}^{\phi}\left(q^2\right) = -iM_A^2\frac{q_\mu q_\nu}{q^2} \tag{11}$$

which adds up to (10) to generate the requisite projector correctly. From these observations one concludes that a vector boson can acquire a mass if it couples to a strictly massless scalar field (See the book by Peskin and Schroeder listed in [1]).

The place where one can find requisite strictly massless scalar fields is given by Goldstone's theorem. This theorem states that there appear strictly massless scalar fields, the Goldstone bosons, for each broken symmetry generator when a continuous symmetry is spontaneously broken, that is, when the minimum energy configuration (ground state) does not respect the symmetry of the Lagrangian. Armed with this theorem, it is now clear that vector boson masses can be naturally generated in gauge theories with spontaneously broken symmetries. A toy model based on a complex scalar $\phi(x)$ and fermion field $\psi(x)$ that realizes aforementioned properties may be taken as

$$\mathcal{L} = -\frac{1}{4}F^{\mu\nu}F_{\mu\nu} + \overline{\psi_L}\left(\not{D} - m_\psi\right)\psi_L + \overline{\psi_R}\left(\not{D} - m_\psi\right)\psi_R + \left[h_\psi\overline{\psi_L}\phi\psi_R + h.c.\right]$$
$$+ \left(D^\mu\phi\right)^\dagger D_\mu\phi - V\left(\phi^\dagger\phi\right) \qquad (12)$$

where h_ψ is the Yukawa coupling and $V\left(\phi^\dagger\phi\right)$ is potential energy density of the scalar field. This Lagrangian represents an Abelian gauge theory invariant under

$$\phi(x) \rightarrow e^{iQ_\phi\alpha(x)}\phi(x),\ \psi_L(x) \rightarrow e^{iQ_{\psi_L}\alpha(x)}\psi_L(x),\ \psi_R(x) \rightarrow e^{iQ_{\psi_R}\alpha(x)}\psi_R(x) \qquad (13)$$

accompanied by transformation of A_μ in (6). For the purpose of illustration the local U(1) invariance is chosen to treat left– and right–chirality fermions differently. The covariant derivatives $D_\mu\phi = \left(\partial_\mu + igQ_\phi A_\mu\right)\phi$, $D_\mu\psi_L = \left(\partial_\mu + igQ_{\psi_L}A_\mu\right)\psi_L$ and $D_\mu\psi_R = \left(\partial_\mu + igQ_{\psi_R}A_\mu\right)\psi_R$ are needed to make derivatives of fields transform as fields themselves under (13). In these expressions Q_ϕ, Q_{ψ_L} and Q_{ψ_R} are, respectively, charges of ϕ, ψ_L and ψ_R under local U(1) symmetry such that $Q_{\psi_R} + Q_\phi - Q_{\psi_L} = 0$ so as to enable the Yukawa interaction ([1,2]).

Focussing on a simple polynomial structure, potential of the scalar field may be taken as

$$V\left(\phi^\dagger\phi\right) = m^2\phi^\dagger\phi + \frac{\lambda}{2}\left(\phi^\dagger\phi\right)^2 \qquad (14)$$

where $\lambda > 0$ and m^2 are model parameters. For $m^2 > 0$ this potential has a minimum at $\phi \equiv \phi^S_{min} = 0$, and an expansion in $\phi - \phi^S_{min}$ reveals a massive scalar interacting with massless A_μ and ψ.

When $m^2 < 0$, that is, when ϕ is a complex tachyon then potential (14) develops a minimum at $\phi \equiv \phi^B_{min}$ where

$$\phi^B_{min} = \left(-\frac{m^2}{\lambda}\right)^{1/2} \qquad (15)$$

with an arbitrary phase reachable by a U(1) transformation. Expanding ϕ around this vacuum expectation value (VEV)

$$\phi(x) = \phi^B_{min} + \frac{1}{\sqrt{2}}\left(h(x) + ig(x)\right) \qquad (16)$$

and using in (12) one finds that particle spectrum now consists of a massive vector boson A_μ, a massive real scalar $h(x)$ and a massive fermion $\psi(x)$ with the masses

$$M_A^2 = 2g^2 \left(\phi_{min}^B\right)^2 , \; m_h^2 = 2\lambda \left(\phi_{min}^B\right)^2 , \; m_\psi = h_\psi \phi_{min}^B \tag{17}$$

where real scalar $g(x)$ in (16) remains strictly massless, and gets swallowed by $A_\mu(x)$ to become massive. In other words, $g(x)$ can be absorbed in $A_\mu(x)$ by a U(1) gauge transformation

$$A_\mu(x) \to A_\mu(x) - \frac{1}{\sqrt{2}} \frac{\partial_\mu g(x)}{g \phi_{min}^B} \tag{18}$$

so that $g(x)$ disappears to generate the longitudinal component of the now-massive vector boson $A_\mu(x)$. This transformation is a reexpression of the fact that a massive vector field, with respect to a massless one such as electromagnetic field, is endowed with a third polarization component parallel to its propagation direction (momentum). (See, for instance, the review volume by Abers and Lee listed in [2]).

In the formulae above, ϕ_{min} stands for the field value for which potential is a minimum, and superscripts S and B designate, respectively, symmetric and broken phases of the U(1) gauge theory at hand. The minimum of the potential at ϕ_{min}^B corresponds to a specific choice for phase of ϕ and thus it breaks or better hides the U(1) invariance. It is in this minimum that gauge boson and fermion acquire their masses.

The toy U(1) gauge theory above illustrates some salient features of mass generation mechanism, that is, the Higgs mechanism. It is not a realistic model of weak interactions; however, it possesses all the features one needs to construct a realistic model. For future reference, one is reminded of the fact that the real scalar field $g(x)$ above is Goldstone boson (of spontaneously broken U(1) invariance), and $h(x)$ is a real field, the so-called Higgs field, that remains in the spectrum as a fingerprint of the Higgs mechanism [1, 2].

3 Electroweak Theory

Armed with the knowledge of Higgs mechanism gained in last section, one can now attempt at constructing a realistic model of weak interactions. A realistic model must, first of all, take into account the fact that gauge bosons to be given mass are electrically charged. Therefore, Abelian models like (12) do not work at all. In fact, a moment of thinking reveals that the simplest gauge group whose gauge bosons consist of a charged sector is SU(2) (the so-called isospin group). This point is best understood by giving the explicit expressions of SU(2) generators (the Pauli matrices)

$$\sigma_3 = \begin{pmatrix} 1 & 0 \\ 0 & -1 \end{pmatrix} , \; \sigma_+ = \frac{1}{\sqrt{2}} (\sigma_1 + i\sigma_2) = \begin{pmatrix} 0 & \sqrt{2} \\ 0 & 0 \end{pmatrix} , \; \sigma_- = \frac{1}{\sqrt{2}} (\sigma_1 - i\sigma_2) = \begin{pmatrix} 0 & 0 \\ \sqrt{2} & 0 \end{pmatrix} \tag{19}$$

with respective gauge bosons A_μ^3, W_μ^+ and W_μ^-. However, it is easy to see that an SU(2)–invariant theory alone is not sufficient for describing Nature.

The reason is that, upon spontaneous symmetry breaking, all these gauge bosons acquire masses, and thus, A^3_μ, the candidate for electromagnetic field, becomes massive. This is an unacceptable result. The way out is to introduce an additional gauge group the simplest of which being a U(1) factor so that gauge boson of this group combines with A^3_μ to form a massless state. The nature of this new U(1) invariance is determined by the Gell-Mann–Nishijima rule

$$Q = I_3 + \frac{Y}{2} \tag{20}$$

which relates electric charge Q to third component of isospin I_3 (eigenvalue of σ_3 above) and hypercharge Y of a particle. Therefore, correct gauge group which describes weak interactions and electromagnetism involves SU(2) and $\mathrm{U}(1)_Y$ symmetries.

At this stage problem is how to arrange matter into SU(2) and $\mathrm{U}(1)_Y$ gauge symmetries. The Yukawa interaction in (12) makes it clear that if both left– and right–handed matter fermions are charged under SU(2) then Higgs field ϕ must be a singlet under both SU(2) and $\mathrm{U}(1)_Y$. This, however, prevents $W^\pm$ bosons to acquire masses – *la raison de entre* for Higgs mechanism and SU(2) invariance. Hence, only matter fields of one handedness, say left-handed ones, must be charged under SU(2). The ones with opposite handedness must be singlets under SU(2) for Higgs field to be an SU(2) doublet to give masses to $W^\pm$ bosons. These observations lead one to the group structure G_{SM} $=\mathrm{SU}(2)_L\otimes\mathrm{U}(1)_Y$ meaning that left-handed fermions are SU(2) doublets, and right-handed ones are SU(2) singlets. In this context, gauge quantum numbers of leptons are

$$\begin{aligned} L_1 &= \begin{pmatrix} \nu_{e_L} \\ e_L \end{pmatrix}_{-1} , \; L_2 = \begin{pmatrix} \nu_{\mu_L} \\ \mu_L \end{pmatrix}_{-1} , \; L_3 = \begin{pmatrix} \nu_{\tau_L} \\ \tau_L \end{pmatrix}_{-1} \\ e_{R_1} &= (e_R)_{-2} \; , \; e_{R_2} = (\mu_R)_{-2} \; , \; e_{R_3} = (\tau_R)_{-2} \end{aligned} \tag{21}$$

and those of quarks are where subscript of each field refers to its hypercharge under the definition (20). On the other hand, charge assignments for quarks are given by

$$\begin{aligned} Q_1 &= \begin{pmatrix} u_L \\ d_L \end{pmatrix}_{1/3} , \; Q_2 = \begin{pmatrix} c_L \\ s_L \end{pmatrix}_{1/3} , \; Q_3 = \begin{pmatrix} t_L \\ b_L \end{pmatrix}_{1/3} \\ u_{R_1} &= (u_R)_{4/3} \; , \; u_{R_2} = (c_R)_{4/3} \; , \; u_{R_3} = (t_R)_{4/3} \\ d_{R_1} &= (d_R)_{-2/3} \; , \; d_{R_2} = (s_R)_{-2/3} \; , \; d_{R_3} = (b_R)_{-2/3} \end{aligned} \tag{22}$$

where subscript of each field refers to its hypercharge under the definition (20). The Higgs field H (which corresponds to the scalar field ϕ in (12) of the toy U(1) model discussed in last section) is given by

$$H = \begin{pmatrix} \varphi^+ \\ \varphi^0 \end{pmatrix}_1 \tag{23}$$

where superscripts on its component fields refer to their electric charges [1,2].

The $SU(2)_L$ and $U(1)_Y$ symmetries refer to electroweak sector which is the main topic of this review volume. Apart from these, as a theory of the constituents of hadrons, that is, quarks and gauge bosons gluing them together, one has to include an additional gauge invariance, $SU(3)_c$, associated with color degrees of freedom. Quarks transform in fundamental of $SU(3)_c$ whose gauge fields are gluons.

Then complete $SU(3)_c \otimes SU(2)_L \otimes U(1)_Y$ invariant SM Lagrangian, explicating all gauge and flavor quantum numbers, is given by [3]

$$\begin{aligned}
\mathcal{L}_{\rm SM} = &-\frac{1}{4} G^{r_a}_{\mu\nu} G^{\mu\nu}_{r_a} - \frac{1}{4} W^{s_a}_{\mu\nu} W^{\mu\nu}_{s_a} - \frac{1}{4} B_{\mu\nu} B^{\mu\nu} \\
&+ \overline{L_i}^{s_f} i \not{D} L^{s_f}_i + \overline{e}_{R_i} i \not{D} e_{R_i} + \overline{Q_i}^{r_f, s_f} i \not{D} Q^{r_f, s_f}_i + \overline{u}^{r_f}_{R_i} i \not{D} u^{r_f}_{R_i} + \overline{d}^{r_f}_{R_i} i \not{D} d^{r_f}_{R_i} \\
&- \Big[(h_u)_{ij} \overline{Q_i}^{r_f, s_f} H^{c, s_f} u^{r_f}_{R_j} + (h_d)_{ij} \overline{Q_i}^{r_f, s_f} H^{s_f} d^{r_f}_{R_j} \\
&+ (h_e)_{ij} \overline{L_i}^{s_f} H^{s_f} e_{R_j} + h.c. \Big] \\
&+ (D_\mu H)^\dagger D^\mu H - m^2 H^\dagger H - \frac{\lambda}{2} \left(H^\dagger H \right)^2
\end{aligned} \tag{24}$$

where $H^c \equiv i\sigma_2 H^\star$ is the charge-conjugate of H, and ranges of various indices are: $r_a = 1, \dots, 8$ for adjoint of $SU(3)_c$, $r_f = 1, \dots, 3$ for fundamental of $SU(3)_c$, $s_a = 1, \dots, 3$ for adjoint of $SU(2)_L$, $s_f = 1, 2$ for fundamental of $SU(2)_L$, $i, j = 1, \dots, 3$ for three flavors of quarks and leptons. The Yukawa matrices are 3×3 complex matrices in the space of fermion flavors. For a generic matter field, including Higgs field H, the gauge-covariant derivative is defined by

$$D_\mu \psi = \left(\partial_\mu - \frac{i}{2} g_s \lambda_{r_a} g^{r_a}_\mu - \frac{i}{2} g_w \sigma_{s_a} W^{s_a}_\mu - \frac{i}{2} g_Y Y_q B_\mu \right) \psi \tag{25}$$

where g^a_μ, W^s_μ and B_μ are gauge fields, $\lambda^{r_a}/2$ (3×3 Gell-Mann matrices), $\sigma^{s_a}/2$ (2×2 Pauli matrices), and Y_ψ (the hypercharge) are generators, and g_s, g_w and g_Y are gauge couplings of $SU(3)_c$, $SU(2)_L$ and $U(1)_Y$, respectively. The field-strength tensors of the gauge fields are given by

$$\begin{aligned}
G^{r_a}_{\mu\nu} &= \partial_\mu g^{r_a}_\nu - \partial_\nu g^{r_a}_\mu + g_s \, f^{r_a bc} g^b_\mu g^c_\nu \\
W^{s_a}_{\mu\nu} &= \partial_\mu W^{s_a}_\nu - \partial_\nu W^{s_a}_\mu + g_w \, \epsilon^{s_a bc} W^b_\mu W^c_\nu \\
B_{\mu\nu} &= \partial_\mu B_\nu - \partial_\nu B_\mu
\end{aligned} \tag{26}$$

where f^{abc} and ϵ^{abc} are, respectively, the structure constants of $SU(3)_c$ and $SU(2)_L$.

The color group $SU(3)_c$ is never broken; quantum chromodynamic (QCD) interactions are (non-)perturbative at (low)high energies – the infamous asymptotic freedom of color group. It is the non-perturbativity of $SU(3)_c$ at the infrared that is responsible for confinement of quarks to form hadrons [4].

As was illustrated for toy U(1) gauge theory in the last section, upon spontaneous breakdown of $SU(2)_L \otimes U(1)_Y$ by the Higgs sector, all fermions acquire masses with a hierarchy determined by their Yukawa matrices:

$$m_u = h_u \frac{v}{\sqrt{2}}\,, \; m_d = h_d \frac{v}{\sqrt{2}}\,, \; m_e = h_e \frac{v}{\sqrt{2}}. \tag{27}$$

where

$$v = \left(-\frac{m^2}{\lambda}\right)^{1/2} \tag{28}$$

is the Higgs VEV. The physical fermion masses arise after diagonalization of these mass matrices. This procedure has no remnant in lepton sector (the flavor structure of h_e is unphysical as long as neutrinos are taken to be massless). However, quark mass matrices, upon diagonalization, leave a physical finger print by dressing all $W^{\pm}\overline{u}^{phys}_{L_i} d^{phys}_{L_j}$ vertices by $(V_{CKM})_{ij}$, the (i,j) entry of the Cabibbo–Kobayashi–Maskawa matrix

$$V_{CKM} = (V_L^u)^{\dagger}\, V_L^d \tag{29}$$

where $V_L^u h_u h_u^{\dagger} (V_L^u)^{\dagger} = \mathbf{1}$ and $V_L^d h_d h_d^{\dagger} (V_L^d)^{\dagger} = \mathbf{1}$ with unitary transformations of the quark fields $u_L^{phys} = V_L^u u$ and $d_L^{phys} = V_L^d d_L$. The unitary matrices $V_R^{u,d}$ that define physical right-handed quarks do not appear in interactions of mass eigenstate fields.

Like fermions, gauge bosons acquire their masses upon $SU(2)_L \otimes U(1)_Y$ breaking. The physical gauge bosons

$$W^{\pm} = \frac{1}{\sqrt{2}}(A^1_{\mu} \mp i A^2_{\mu})\,, \; Z_{\mu} = \frac{g_w A^3_{\mu} - g_Y B_{\mu}}{\sqrt{g_w^2 + g_Y^2}}\,, \; A_{\mu} = \frac{g_w W^3_{\mu} + g_Y B_{\mu}}{\sqrt{g_w^2 + g_Y^2}} \tag{30}$$

obtain masses $M_W = \frac{1}{2} g_w v$, $M_Z = \frac{1}{2}\sqrt{g_w^2 + g_Y^2}\, v$ and $M_A = 0$. The model correctly predicts massive charged bosons $W^{\pm}$ plus a massless gauge boson, A_{μ}, corresponding to electromagnetic field. However, it also predicts a neutral massive boson Z_{μ} experimental confirmation of which forms an important test of electroweak theory. At this point it may be instructive to determine cost of giving mass to gauge bosons. Before electroweak breaking (corresponding to ϕ^S_{min} in last section), all gauge bosons W^i_{μ} and B_{μ} are massless, and the Higgs doublet (23) consists of four real fields $\varphi^{\pm}$, $\Re[\varphi^0 - v]$ and $\Im[\varphi^0 - v]$. In course of spontaneous breakdown of $SU(2)_L \otimes U(1)_Y$ (corresponding to ϕ^B_{min} of last section), $\varphi^{\pm}$ are spent for giving mass to $W^{\pm}$ bosons, and $\Im[\varphi^0 - v]$ is spent for giving mass to Z boson. Hence, three Goldstone bosons are spent for giving mass to gauge bosons. The real part of neutral component of Higgs field, $\Re[\varphi^0 - v]$, is nothing but the Higgs boson – an important finger print of spontaneous breakdown of $SU(2)_L \otimes U(1)_Y$. Its mass-squared

$$m_h^2 = 2\lambda v^2 = -2m^2 \tag{31}$$

is a direct probe of the tachyonic mass of the Higgs field H [2].

Since their first observation at CERN, the $W^\pm$ and Z boson masses have been measured with increasing precision:

$$M_W = 80.403\,\text{GeV}\,,\ M_Z = 91.1876\,\text{GeV} \tag{32}$$

as listed by PDG [5]. Continuing with PDG tables, the fermion masses read as

$$\begin{aligned}
&m_e = 0.511 \text{ MeV}\,,\ m_\mu = 105.658 \text{ MeV}\,,\ m_\tau = 1.777 \text{ GeV}\\
&m_u = 1.5 - 3 \text{ MeV}\,,\ m_c = 1.25 \text{ GeV}\,,\ m_t = 174.2 \text{ GeV}\\
&m_d = 3 - 7 \text{ MeV}\,,\ m_s = 95 \text{ MeV}\,,\ m_b = 4.2 \text{ GeV}
\end{aligned} \tag{33}$$

where u, d, s quark masses are $\overline{\text{MS}}$ masses evaluated at 2 GeV scale. The quark mixings, parameterized by V_{CKM}, have been measured and will continue to be measured with increasing precision via experiments on various B meson decays (the b quark is heavy enough to apply perturbative QCD).

Consequently, in the SM the only particle which has not yet been observed is the Higgs boson h, and its experimental discovery and theoretical understanding is one of the hottest topics of high energy physics today [1, 2, 6]. From theoretical perspective, the model itself does not provide a means of computing m_h in terms of known particle masses. There are certain physical requirements, however, which determine likely ranges of m_h. An important constraint in the class is unitarity which requires cross section for a scattering process to behave as $1/\text{energy}^2$ at high energies. The importance of unitarity constraint is best seen by examining scattering of massive vector bosons. Indeed, scattering of longitudinal $W^\pm/Z$ (massive vector bosons, with respect to photon, develop a third polarization component parallel to their momentum) grow with their momenta. More explicitly, for $s \gg M_W^2$,

$$\mathcal{A}\left(W_a^+ W_b^- \to W_c^+ W_d^-\right) = \frac{1}{v^2}\left(s + t - \frac{s^2}{s - m_h^2} - \frac{t^2}{t - m_h^2}\right) \tag{34}$$

where Mandelstam variables are defined by $s = (p_a + p_b)^2$ and $t = (p_a - p_c)^2$. Clearly, this scattering amplitude must approach to that of charged Goldstone bosons contained in (23) at high energies, and hence, it can cause violation of unitarity. This yields an upper bound of $m_h \lesssim 900$ GeV. One arrives at similar bounds via triviality (related to renormalization group flow of the quartic coupling in Higgs potential) or vacuum stability arguments. Hence, one expects Higgs boson to weigh below TeV scale for Higgs mechanism to account for observed mass spectra of matter and force carriers. In fact, global fits to various electroweak precision observables, including gauge boson masses and several asymmetries, agree with the SM predictions. These fits require Higgs boson to weigh below 260 GeV at 95% C. L. [2].

Higgs searches at various experiments have given negative results. The LEP II experiment at CERN, e^+e^- collisions at $\sqrt{s} = 209$ GeV, has ended by reporting the lower bound [7]

$$m_h \geq 114.4\,\text{GeV} \tag{35}$$

at 90% C. L. Search for Higgs boson will continue at next generation colliders: the Large Hadron Collider (LHC) at CERN, the International Linear Collider (ILC), CLIC, and others. In each experiment, collisions of two beams of hadrons/leptons generates Higgs boson in certain channels. The question of what channel is more viable for Higgs discovery depends exclusively on the Higgs mass since couplings of the Higgs boson to itself and other matter species are determined entirely by its mass alone [6]:

$$g_{hff} = \frac{m_f}{v}\,,\; g_{hVV} = 2\frac{M_V^2}{v}\,,\; g_{hhVV} = 2\frac{M_V^2}{v^2}\,,\; g_{hhh} = 3\frac{m_h^2}{v}\,,\; g_{hhhh} = 3\frac{m_h^2}{v^2} \tag{36}$$

where $V = (W^{\pm}, Z)$ and f is a generic fermion. The relations among certain couplings reveal hidden $S(2)_L \otimes U(1)_Y$ symmetry. The SM Lagrangian (24), being a renormalizable quantum field theory, has all the ingredients for inclusion of quantum theoretic effects in Higgs (and any other field in the spectrum) boson production and decays. These corrections, which are crucial for high-energy high-intensity collider processes, will not be detailed in this review volume [2].

4 Collider Searches for Higgs Boson

In this section we shall discuss, somewhat briefly, Higgs boson search strategies at high-energy colliders, in particular, LHC. (See the excellent book [6] for a detailed analysis of Higgs boson search.) The experiments at other colliders, especially at ILC as a precision machine, are certainly important; however, as the earliest experiment to start, we focus on the LHC signals here.

The LHC experiment (see the URL: `http://greybook.cern.ch/` as well as [5]), which is expected to start running this year, is a pp collider where each proton beam has 7 TeV energy. As for any collision process, a crucial parameter is luminosity L which relates rate of occurrence, R, of a given event to its interaction cross section σ_{int} via $R = L\sigma_{int}$. The luminosity is a measure of how frequent the beams of particles are showering on the interaction point:

$$L = \frac{\nu}{4\pi} \frac{n_1 n_2}{\sigma_{hor}\sigma_{ver}} \tag{37}$$

where n_1 and n_2 are numbers of particles in bunches (forming the beam), ν frequency with which the bunches collide, σ_{hor} and σ_{ver} characterize Gaussian beam spreads in transverse plane (in horizontal and vertical directions). The accelerator parameters at the LHC are such that planned luminosity is 10^{34} cm^{-2}s^{-1}.

The LHC complex consists of four detectors: ATLAS and CMS (general purpose experiments for new physics search), ALICE (a special experiment for quark-gluon plasma search), and LHCb (a special experiment for measurements of flavor and CP violation effects in B meson system). There are other experiments as well: the LHCf experiment which will examine high-rapidity region for laboratory generation of ultra-high energy cosmic rays, and TOTEM experiment which will measure cross sections and diffractive processes in conjunction with the CMS detector.

In pp collisions, production and detection of the Higgs signal proceed via different channels at different Higgs mass values and luminosities. The fundamentally relevant processes are

$$\begin{aligned}
&\text{Gluon fusion: } gg \to h \to \gamma\gamma\,, VV^\star \\
&\text{Weak boson fusion: } qq \to qqV^\star V^\star \to qqh \\
&\text{Drell-Yan: } q\overline{q} \to hV^\star \\
&\text{Radiation off top: } gg, q\overline{q} \to ht\overline{t}
\end{aligned} \tag{38}$$

each of which can be estimated by using (36) and contrasted with experimental data available then. Clearly, gluon fusion $g\,g \to h$ and diphoton decay $h \to \gamma\gamma$ arise at the loop level. In all these channels, except for gluon fusion, detection of the Higgs boson requires reconstruction of its mass and couplings from decay products.

Figure 1 depicts integrated luminosity needed to discover a Higgs boson of given mass at the LHC. The figure treats production channels one by one, and it provides requisite information for determining what channel dominates at what values of Higgs boson masses for an anticipated integrated luminosity in units of fb^{-1}.

From the figure it is clear that an integrated luminosity of 100 fb^{-1} suffices to discover a Higgs boson as heavy as 600 GeV (which is well inside the range expected from unitarity and other arguments). In case LHC achieves this size of luminosity it will certainly discover a Higgs boson within 5 σ accuracy. The largest contribution among all channels comes from $h \to Z^\star Z^\star \to 4\ell$ for $m_h \sim 2M_Z$. For small m_h, close to LEP II lower bound, a 30 fb^{-1} luminosity is sufficient to detect Higgs boson via gluon fusion or diphoton decay or $WW^\star$ production. For high-mass Higgs, $m_h \gtrsim 300$ GeV, $h \to ZZ$ dominates. Irrespective of what channel is actually dominating, LHC will certainly discover a Higgs boson within 5 σ accuracy if electroweak breaking is the correct mechanism for generating mass of matter.

5 Summary and Outlook

In this brief review we have provided a pedagogical introduction into 'the problem of mass generation' by putting emphasis on connection between mass and symmetry properties of a field. Then we have proceeded to explain basic mechanism – the Higgs mechanism – by a toy U(1) gauge theory. Afterwards, we have mentioned all the reasonings and constraints for constructing a realistic theory of weak interactions. This led us to a theory of electroweak interactions. Finally, we have given a brief discussion of LHC as a collider and Higgs

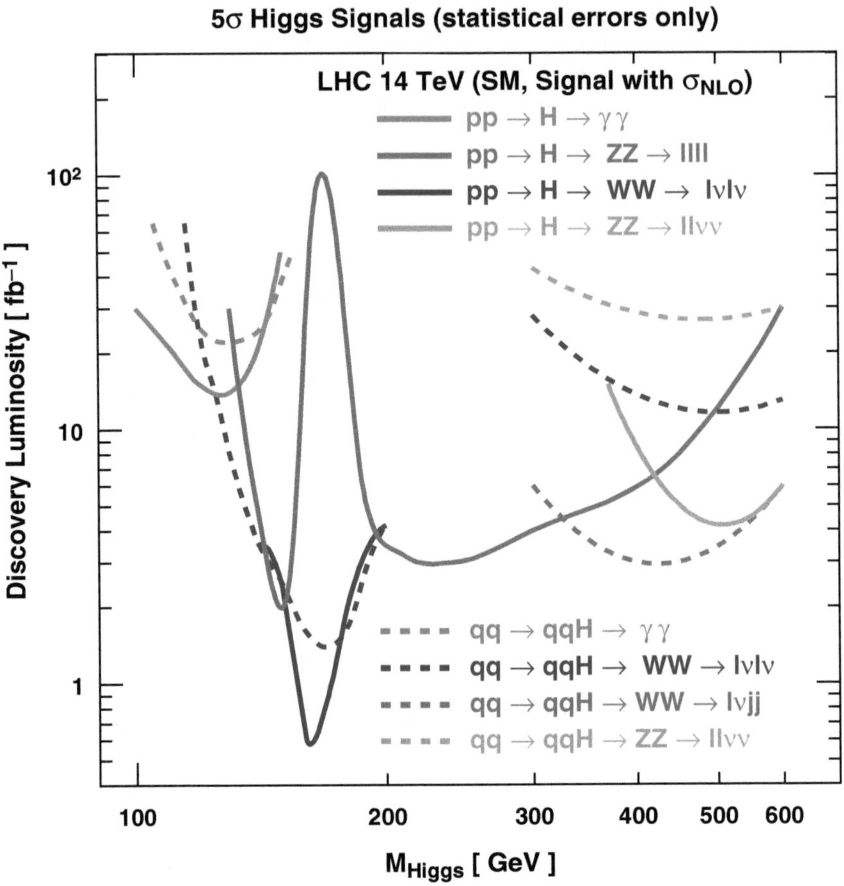

Fig. 1. 5σ discovery luminosity vs. Higgs boson mass at the LHC. The figure is taken from [8].

boson searches therein. The LHC, if reaches a luminosity $\mathcal{O}(100\ \text{fb}^{-1})$ will be capable of discovering a Higgs boson within the mass range indicated by unitarity arguments.

The SM, however, hardly stands as a complete theory of Nature above the electroweak scale. The reason is that the Higgs sector develops a strong sensitivity to ultraviolet (UV) cut-off once effects of quantum fluctuations are included. More explicitly,

$$m_h^{2\,corr} = m_h^2 + \frac{3\Lambda^2}{8\pi v^2}\left[m_h^2 + 2M_W^2 + M_Z^2 - 4m_t^2\right] \tag{39}$$

where Λ is the UV cut-off, m_h is tree-level (classical) Higgs mass defined in (31), and $m_h^{2\,corr}$ is the one-loop (quantal) Higgs boson mass. This divergence is quadratic, not logarithmic, and hence it poses a serious problem since larger the Λ (the highest energy scale up to which SM works well) larger the quantal correction. For instance, if SM is a theory of fundamental particles and interactions among them up to the gravitational scale $\Lambda = (8\pi G_N)^{-1/2}$ then quantum correction turns out to be some 16 orders of magnitude larger than the tree-level Higgs mass m_h (which is expected to lie at the electroweak scale). One way of evading this is to fine-tune quantum correction to zero which requires $m_h \sim 320$ GeV. This requirement involves a huge fine-tuning, however. A more natural way out of this impasse should soften dependence on Λ or should give a physical meaning to Λ beyond what one might know of in the SM.

The way out is to embed SM into another theoretical framework which exhibits good UV behavior. One such playground is provided by 'supersymmetry' which controls Higgs mass just like chiral symmetry controls the fermion masses. It doubles the particle spectrum of SM by introducing a fermion for each boson and vice versa so that loops of a particle and its supersymmetry partner cancel out in accord with Spin-Statistics Theorem (for a pedagogical introduction to supersymmetry, see [9]).

Other than supersymmetry, a higher dimensional spacetime whose gravitational interactions possess a Newton constant $\mathcal{O}$(TeV) is known to bring an alternative solution to UV sickness of the SM Higgs sector. This framework, large extra dimensions, promotes electroweak scale TeV to be a fundamental constant of Nature, and treats Newton constant in four dimensions as a derived quantity (for a pedagogical introduction to extra dimensions, see [10]).

A beginner having sufficient 'theory minimum' about quantum field theory is expected to follow observations and arguments given in this review volume.

6 Acknowledgements

Author gratefully acknowledges organizers of International Summer School on High Energy Physics at Akyaka, especially Meltem Serin for her hospitality and efforts to make the school an excellent one. Author thanks participants of the school, specially Namik K. Pak and Goran Senjanovic, for fruitful discussions and conversations. This work was partially supported by Turkish Academy of Sciences through GEBIP grant, and by the Scientific and Technical Research Council of Turkey through project 104T503.

References

1. T.P. Cheng and L.F. Li, *Gauge theory of Elementary Particle Physics*, Clarendon Press, 1984; C. Itzykson and J.B. Zuber, *Quantum Field Theory*, Mac Graw Hill, 1985; D. Bailin and A. Love, *Introduction to Gauge Field Theory*, Adam Hilger, 1986; M.E. Peskin and D.V. Schroeder, *An introduction to Quantum Field Theory*, Addison-Wesley, 1995.

2. E. Abers and B. Lee, Phys. Rep. 9 (1973) 1; M.A.B. Beg and A. Sirlin, Phys. Rept. 88 (1982) 1; M.E. Peskin, hep-ph/9705479; S. Dawson, hep-ph/9901280; M. Spira and P.M. Zerwas, hep-ph/9803257; C. Quigg, hep-ph/9905369; M. Carena and H.E. Haber, Prog. Part. Nucl. Phys. **50**, 63 (2003) [arXiv:hep-ph/0208209]; A. Djouadi, arXiv:hep-ph/0503172.
3. S. Glashow, Nucl. Phys. 22 (1961) 579; S. Weinberg, Phys. Rev. Lett. 19 (1967) 1264; A. Salam, in "Elementary Particle Theory", ed. N. Svartholm, Almqvist and Wiksells, Stockholm (1969) p. 367.
4. W.J. Marciano and H. Pagels, Phys. Rept. **36**, 137 (1978); V.A. Novikov, L.B. Okun, M.A. Shifman, A.I. Vainshtein, M.B. Voloshin and V.I. Zakharov, Phys. Rept. **41**, 1 (1978); E. Reya, Phys. Rept. **69**, 195 (1981).
5. Particle Data Group, K. Hagiwara *et al.*, Phys. Rev. D66 (2002) 010001; S. Eidelman *et al.*, Phys. Let. B592 (2004) 1.
6. J.F. Gunion, H.E. Haber, G.L. Kane and S. Dawson, *The Higgs Hunter's Guide*, Westview Press; New Ed. 2001.
7. The LEP Collaborations and LEP Higgs Working Group, ALEPH-2000-074 CONF 200-051.
8. M. Dittmar and S. Nicollerat, CMS–Note 2001-036 (See the URL: `http://cmsdoc.cern.ch/documents/01/note01-036.pdf`).
9. H.P. Nilles, Phys. Rept. **110**, 1 (1984); M. E. Peskin, *Supersymmetry: Theory*, World Scientific, 1996; S. P. Martin, arXiv:hep-ph/9709356.
10. N. Arkani-Hamed, Talk at SLAC Summer Institute, 1998 (See the URL: `http://www.slac.stanford.edu/gen/meeting/ssi/1998/arkani.html`); C. Csaki, J. Hubisz and P. Meade, arXiv:hep-ph/0510275.

Heavy Flavor and B Physics

George W.S. Hou

Department of Physics, National Taiwan University, Taipei, Taiwan 10617, R.O.C.
wshou@phys.ntu.edu.tw

This note is based on several lectures on B physics, given to unspecialized students in particle physics. The physics of flavor is presented from an experimental point of view. The theme is CPV measurement and the CKM matrix. Some recent topics are also discussed, with a projection into the future.

1 Setting the Scene

With the riddle of the nucleus resolved by the neutron discovery and the pion proposal, isospin symmetry, or $m_p \cong m_n$, was an early harbinger of "flavor", which is now understood as due to m_u, $m_d \ll m_N$. The $\pi \to \mu \to e$ decay chain, discovered in 1947 using emulsions developed during WWII, gave us the muon, which is "*just like the electron*", but 200 times heavier. I.I. Rabi's question: "Who ordered that?", marks the beginning of flavor physics.

The muon is the template of 3-body tree level weak decay ($x = m_e^2/m_\mu^2$),

$$\Gamma_\mu = \frac{G_F^2 m_\mu^5}{192\pi^3}(1 - 8x + 8x^3 - x^4 - 12x^2 \ln x). \tag{1}$$

Loop induced $\mu \to e\gamma$ and $g-2$ processes are still of great interest.

1.1 Kaon: the "Strange" Template

Repeating the muon discovery, "strangeness" became the template for flavor.

Long Lifetime With $m_K^2/m_\pi^2 \sim 13$ large, it was hard to understand the smallness of $\Gamma_K/\Gamma_\pi \sim 2$. This "strangeness" lead to the Cabibbo proposal. In quark language it states that one has $d_C = d\cos\theta_C + s\sin\theta_C$ in weak interactions, with the Cabibbo angle $\sin\theta_C \simeq 0.22$ suppressing strangeness decay.

K^0-$\bar{K}^0$ Mixing With strangeness, K^0 and $\bar{K}^0$ are antiparticles of each other, and eigenstates of the strong interaction, but can mix via the weak interaction into the CP even and odd K_1^0 and K_2^0 mesons, where C and P are charge conjugation and parity operations. With parity violation, it was believed that CP remained conserved, then $K_1 \to \pi\pi$ and $K_2 \to \pi\pi\pi$ but not vice versa. With $3m_\pi \sim m_K$, one gets $\tau_{K_1} \ll \tau_{K_2}$, agreeing with data, and one observes the so-called mixing, or oscillation in time, of $K^0 \leftrightarrow \bar{K}^0$.

Suppressed SCNC Neutral current induced $s \to d$ effects were expected at $\sim\sin\theta_C \cos\theta_C$ of charged current effects. One then can not understand the slowness of K^0–$\bar{K}^0$ oscillations, while $K_2 \to \mu^+\mu^-$ was also highly suppressed. Glashow, Iliopoulos and Maiani (GIM) proposed adding a second weak doublet $s_C = -d\sin\theta_C + s\cos\theta_C$ to the Cabibbo proposal. This completes a 2×2 rotation in charge $-1/3$ flavor space. Miraculously, this not only removes tree level Strangeness Changing Neutral Currents (SCNS), it renders the second order weak box diagram calculable, with both u and c (partner of s in the new doublet) quarks in the loop, allowing the charm mass to be estimated.

CP Violation In 1964 it was discovered that the "K_2^0" meson actually decayed to $\pi\pi$ at $\sim1/500$, which can be parameterized by an admixture of $|\varepsilon_K| \sim 2\times10^{-3}$ of K_1 in "K_2". The mass eigenstates were therefore called K_S and K_L, for short and long lived, respectively. CPV occurred in K^0-$\bar{K}^0$ mixing.

Digression on CPV

The invention of the GIM mechanism, together with the challenge to understand CPV, lead to a further observation. Kobayashi and Maskawa pointed out that, by enlarging the 2×2 rotation matrix to a 3×3 unitary matrix, the gauge theory of weak interactions has a *unique*, irremovable phase that is CP violating. CPV called for the existence of 3 generations.

C maps particles to their antiparticles. Combining with P, the CP operation maps left-handed fermions to right-handed antifermions. Given that particle and antiparticle are on equal footing in dynamical theories such as QED (and QCD), it was hard to understand why the Universe seems to be made only of "particles", p, n and e. Where have all the antimatter gone? To address this pressing question, and motivated by the discovery of CPV, Sakharov formulated the 3 conditions for the Matter Universe, i.e.

1) baryon number violation, 2) CPV, and 3) out of equilibrium.

However, the CPV observed in ε_K within the KM model, is orders of magnitude too small. This adds to the importance of continued search for CPV.

It is illuminating to consider CPV in the quantum mechanical context. The essence of quantum, or wave, behavior is the interference of *complex* amplitudes, with a common i in the time evolution of quantum states. From atomic to condensed matter physics, the dynamics is nothing but Maxwell theory, and the "reality" of the coupling constant e is taken for granted. We now understand that gauge charge is real.

Complexity in dynamics, the existence of *complex* couplings, such as the Yukawa couplings in the Standard Model (SM), makes quantum interference much more interesting — which is CPV phenomena.

1.2 Heavy Flavor: $c/\tau/b/t$

The story of heavy flavor starts with the discovery of charm in 1974, and culminates in the top discovery in 1995 that completed the 3rd generation.

On the opposite end, the neutrinos were recently found to have mass [1], but not more than eV scale (and probably far less). In between we have the familiar $u/d/s$ quarks and e, μ charged leptons. With QM underlying the periodic table, and QCD underlying the proliferation of hadrons, we do not yet know whether anything underlies the "spectroscopy of flavor".

Charm

The GIM mechanism required the existence of the "charm" quark, even an estimate of m_c from K^0-$\bar{K}^0$ mixing. But this does not take away any glory from the experimental discovery of charm, first in its hidden form as J/ψ.

The discovery of J in $p + \text{Be} \to e^+e^- + X$ at BNL, and ψ in $e^+e^- \to$ hadrons and leptons at SLAC, caused the "November Revolution": a new very heavy hadron with very narrow width. The narrowness can be qualitatively accounted for by invoking the OZI rule, which implied a new "flavor" quantum number beyond strangeness. The J/ψ discovery precipitated the acceptance of quarks, acceptance of QCD as the underlying strong dynamics, and the existence of a new spectroscopy. The Standard Model became adopted.

The SPEAR e^+e^- collider with the solenoidal MARK I detector marked the start of e^+e^- collider physics. The D mesons, or "open charm" ($c\bar{q}$ mesons) as compared to the "hidden charm" ($c\bar{c}$ mesons called charmonium), followed in 1975. The resonance $\psi(3770)$, above $D\overline{D}$ threshold, decayed dominantly into $D\overline{D}$ meson pairs without OZI violation. Spectroscopic activities have been renewed recently, after a 25 year hiatus, by the discovery of new narrow states such as the $X(3872)$ [2] and $D_{sJ}(2317)$ [3].

D^* mesons, analogues of the ρ and K^* mesons, further confirmed the quark model picture, with $D^{*0} \to D^0\pi^0$, $D^0\gamma$ while $D^{*+} \to D^0\pi^+$, $D^+\pi^0$, $D^+\gamma$ understood from kinematics. The pseudoscalar mesons, the lightest charmed particles, can decay only weakly. It was found that $\tau_{D^+}/\tau_{D^0} \sim 2.5$, indicating that hadronic effects are important for D meson weak decay.

We shall not dwell much on charm physics, except a few special issues.

Tau

Although the KM picture for CPV anticipated the existence of the 3rd generation, there is no prediction for their masses. The discovery of the τ lepton followed swiftly after charm, however, in 1975 at SLAC, with the help of theoretical calculations that generalized from Eq. (1). From $\tau \to e\nu\bar{\nu}$ and $\mu\nu\bar{\nu}$ decay, one has the signature of $e^+e^- \to \tau^+\tau^- \to e^\pm\mu^\mp +$ missing energy. The process $e^+e^- \to \tau^+\tau^-$ has been studied from SPEAR up to LEP energies, confirming SM couplings of the τ lepton to great precision.

Again, we shall not cover tau physics except a couple of rare decays.

Beauty

The discovery of the bottom quark, or "beauty" flavor, repeated the J/ψ: the hidden beauty ($b\bar{b}$) Υ meson was discovered in 1977 at Fermilab as a bump in $m_{\mu\mu}$ in $p + \text{Be} \to \mu^+\mu^- + X$ events. This stimulated the development of the e^+e^- colliders CESR at Cornell, and DORIS at DESY, capable of running around 10 GeV center of mass (CM) energies. By 1980, the CLEO detector operating at CESR found the $\Upsilon(4S)$ resonance, which opened the way for open beauty meson studies. The B mesons ($b\bar{q}$) were discovered in 1983, while the "bottomonium" spectroscopy further confirms the quark model and the underlying QCD. With $\tau_{B^+}/\tau_{B^0} \sim 1.076$, hadronic effects are far less important for B mesons than for D mesons.

Top

With τ and b discovered, the charge $+2/3$ quark seemed just around the corner. But the top quark did not appear until 1995. At 174 GeV, it is uniquely heavy among fermions. Because of this, the top quark as a probe of flavor has yet to start, so we shall also not cover it.

1.3 The Centrality of B

The $c/\tau/t$ lifetimes are unsuppressed, hence loop induced decays are much more subdued. In contrast, B physics is central to flavor and CPV, because it repeats the pattern seen in strangeness, and is in fact far richer.

We retrace the discoveries, highlight the physics, and point out the techniques that became crucial for e^+e^- and hadronic "B factories".

Long B Lifetime and CKM Pattern

Vertex detectors with sub-mm accuracy, aimed at c/τ lifetime measurement, discovered the long τ_b in 1983. The MAC and MARK II experiments [4] at PEP observed large impact parameter events (high p_T leptons with respect to the jet axis), implying that b hadron lifetime was at ps level, several hundred times longer than scaling up from weak decay of the τ. This implied that V_{cb} and V_{ub} are much smaller than $\lambda \equiv V_{us} \sim 0.22$ and was not anticipated.

Reports in summer 1983 on semileptonic b decay indicated [5] $b \to c\ell\nu$ dominance, i.e. the E_ℓ spectrum cuts off with a recoil system compatible with the D meson, but no higher E_ℓ component corresponding to $b \to u\ell\nu$. This was the first clear indication that $|V_{ub}| \ll |V_{cb}|$. We now know that [6]

$$V_{us} \equiv \lambda \sim 0.22,\ V_{cb} \equiv A\lambda^2 \simeq 0.04,\ |V_{ub}| \equiv A\lambda^3\sqrt{\rho^2+\eta^2} \sim 0.004. \quad (2)$$

Large B^0 Mixing and Heavy Top

"Accompanying" the SN1987A supernova, the ARGUS experiment announced the discovery of rather large B^0-$\bar{B}^0$ mixing [7]. Like sign dilepton events with $\sim$100 pb^{-1} data at the DORIS storage ring running on the $\Upsilon(4S)$ gave the mixing parameter $r = 0.21 \pm 0.08$, just below a preceding limit given in 1986 by the CLEO experiment, and the confirmation from CLEO came in 1989. The $x_B = \Delta m_B/\Gamma_B \sim 1$ value is in contrast to the prejudice that $x_B \ll x_K$, which was due to the prevailing expectation, since 1977, that the top quark was "just around the corner". A flourish of papers followed to claim a heavy top, $m_t > 50$ GeV. It took another decade for the top quark to appear.

Rare B Decays: $B \to K^*\gamma$ and $X_s\gamma$

In 1993, the CLEO experiment observed 13 $B \to K^*\gamma$ events with roughly 1.4 fb^{-1} data, and measured the radiative decay at 4–5$\times10^{-5}$ [8]. This was the first evidence for $b \to s\gamma$, forbidden within the SM at the tree level by gauge invariance. The measurement came after a major investment on CsI(Tℓ) crystal electromagnetic calorimeter (ECAL). By end of 1994, CLEO observed [8] the *inclusive* $b \to s\gamma$ process, by partially reconstructing X_s via $K + n\pi$, with n up to 4, and at most one π^0. The $B \to X_s\gamma$ rate, identified with $b \to s\gamma$, was found to be 2–3$\times10^{-4}$, in agreement with SM expectations enhanced by QCD. The m_{X_s} spectrum from partial reconstruction shows a peak consistent with K^* at low mass, and resonances at higher mass.

The rare $b \to s\gamma$ decay is mediated by "penguin" loops, with real photon emission. The b quark first "decays" to a virtual W and t pair. With the emission of the photon, the virtual Wt pair recombines into an s quark. The photon couples to the $b \to s$ transition current via an effective $\sigma_{\mu\nu}$ dipole, and at leading order in strong interactions, the process is highly GIM suppressed. But, when one considers QCD corrections, the GIM suppression is circumvented at next order [9]. Thus, though higher order, because of extreme suppression of the leading term, the rate becomes greatly enhanced at α_s order. The CLEO measurement confirms this picture. One outcome of this measurement is the most stringent bound on the charged Higgs boson mass, $m_{H^+} > 320$ GeV [6], by replacing W^+ by H^+ in the loop [10].

With the improved CLEO II, the ARGUS experiment came to an end.

Towards B Factories

With long B lifetime, large B^0-$\overline{B}^0$ mixing and rare "BCNC" decays, one anticipates further "repeat" of CPV phenomena in the B system. It turned out to be far richer than the kaon system. The mixing-decay interference mechanism was proposed (Lecture 2), which received great boost from the observation of large B mixing in 1987. Before long, there was a world competition for "B factories", to observe CPV in B system.

The KEK and SLAC proposals picked up the idea [11] of an asymmetric e^+e^- collider. With a boosted B meson, new silicon vertex detectors allowed one access to mixing-decay interference measurement. Construction of KEKB and PEP-II started in the first half of the 1990s.

In 1992, the CDF experiment at the Tevatron fully reconstructed $\sim$10 $B^\pm \to J/\psi K^\pm$ events [12], demonstrating that B physics at hadron colliders is feasible. With the advantage in cross section, and techniques such as vertexing, forward detector concepts were advanced to further utilize a larger boost and allow more space (especially for particle identification, or PID, and ECAL). The LHCb detector will run in 2008 at LHC. Unfortunately, the BTeV detector, proposed for the Tevatron, was canceled in 2005.

1.4 The B Factories

The KEKB accelerator, operating with 3.5 e^+ on 8.5 GeV e^-, produces the $\Upsilon(4S)$ resonance with $\beta\gamma = 0.425$, while PEP-II has $\beta\gamma = 0.56$. The main difference between the two is in beam crossing at the Interaction Region (IR). KEKB uses finite angle crossing, while PEP-II uses the more conservative zero angle crossing, with beam separation by permanent magnets at IR.

PEP-II took the initial lead after 1999 commissioning, but was overtaken by KEKB in 2001. KEKB now has almost twice the accumulated luminosity, and is the "brightest" collider in the world. Between the two, a total of over 1000 fb^{-1} (1 ab^{-1}) data has been logged so far. Note that, at $\Upsilon(4S)$ energy

$$1\ \text{fb}^{-1} \cong 1.12\,\text{M}\ B\overline{B}\ \text{pairs}, \tag{3}$$

together with about 1.3M (0.89M) $c\bar{c}$ quark ($\tau^+\tau^-$) pairs, so a B factory is also a τ/charm factory. With $\Upsilon(4S) \to B\overline{B} \cong 100\%$ of the time, and the closeness of B^+ and B^0 in mass, the amount of B^+B^- and $B^0\bar{B}^0$ are almost the same, and the fraction is measurable. With $\Upsilon(4S)$ a $J = 1$ resonance, the $B\overline{B}$ are flavor "locked", or "entangled" at production. This fact, that a B decay at time t (in CM frame) fixes the $\overline{B}$ flavor of the other B at same time, is utilized in the mixing-decay interference study. The quantum entanglement effect has been recently studied and confirmed by Belle [13].

The Belle (see Fig. 1) and BaBar detectors have similar design concepts. From inside out: vertex detector, drift chamber, PID, ECAL, solenoid (1.5 Tesla), and flux return instrumented to detect muons and K_L. The main difference is in PID: Belle uses Aerogel Cherenkov Counters (plus TOF), a threshold device, while BaBar uses a quartz bar to extract Cherenkov light by total internal reflection, projecting into a water tank at the backside. Other differences are the 3 (later 4) layers vs 5 layers of Si for the vertex detector, and a closer (and no backward endcap) ECAL for Babar.

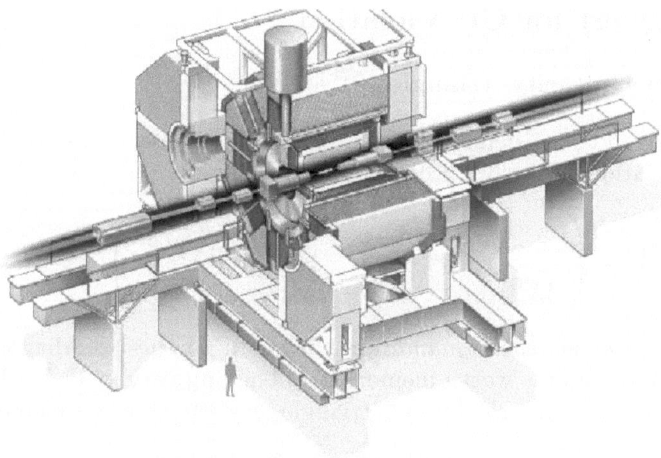

Fig. 1. The Belle detector.

Let us briefly summarize detection requirements for each subdetector. The SVD should have sufficient resolution for Δz along the boost direction, which translates to Δt of the two B mesons. It must survive (longevity) and operate (tolerance) in a high radiation environment. The Belle SVD2 upgrade in 2003 improved the 1 MRad radiation tolerance of SVD1 to better than 20 MRad, changed from 3 to 4 layers, and with improved angular coverage. Together with a smaller 1.5 cm (vs 2 cm) beampipe radius, the IP (impact parameter) resolution was improved by over 60%. Currently the SVD2 data fraction is already much larger than the original 152M $B\overline{B}$ accumulated with SVD1.

The hadron PID, i.e. charged $\pi/K/p$ separation, is as important for CPV studies as vertexing. The lower momentum range is used for kaon tagging, while for two-body decay modes one needs $\pi^\pm/K^\pm$ separation up to 4.4 GeV (for Belle) in forward direction. A complication is operating phototubes in a strong B field. The Belle PID system consists of dE/dx from CDC, TOF, and the ACC. The ACC is based on hydrophobic silica aerogel of ultra low density, with $n = 1.01$–1.028 in barrel, to 1.03 in forward endcap. The DIRC detector of BaBar provides more information with lower pion fake rate.

The ECAL (called ECL at Belle) measures shower energy of e, γ and π^0, gives best possible energy and position resolution, and optimal angular and energy coverage. Matching with tracker gives electron ID. As shown by CLEO II, the expense on crystals (thallium doped CsI) cannot be spared, but the cost, and the schedule for production, are limiting factors.

The low energy handicaps hadron calorimetry, but the iron flux return is instrumented (by layers of RPCs, or resistive plate chambers) to detect muons (>800 MeV) by pairing with tracker. It also provides partial detection of K_L mesons that largely penetrate all inner detectors, by pairing with ECL response. At Belle, this subdetector is called the KLM (K_L/Muon).

2 The Quest for CP Violation

2.1 CKM Unitarity Triangle

With λ, A and ρ, η defined in Eq. (2), it is useful to give the Wolfenstein parametrization of the 3×3 quark mixing matrix,

$$V = \begin{pmatrix} 1-\frac{1}{2}\lambda^2 & \lambda & A\lambda^3(\rho - i\eta) \\ -\lambda & 1-\frac{1}{2}\lambda^2 & A\lambda^2 \\ A\lambda^3(1-\rho-i\eta) & -A\lambda^2 & 1 \end{pmatrix} + \mathcal{O}(\lambda^4), \tag{4}$$

which puts the *complex dynamics* in the small, farthest off-diagonal element of V_{ub} and V_{td}, and is worth memorizing. The unitarity of V, i.e. $V^\dagger V = I$, is best visualized geometrically as a triangle, e.g. the $(VV^\dagger)_{db}$ element is

$$V_{ud}V_{ub}^* + V_{cd}V_{cb}^* + V_{td}V_{tb}^* = 0, \tag{5}$$

which is illustrated in Fig. 2. To over-constrain its angles and sides is the central theme at the B factories. The phase angles ϕ_1, ϕ_2, ϕ_3, used by Belle are alternatively called β, α and γ by BaBar.

Tree Level B Decays

The inclusive (quark level) diagram is given in Fig. 3(a), where we have suppressed the virtual W. The leading decays are $b \to c\bar{u}d$, $c\bar{c}s$, and the semileptonic $b \to ce\bar{\nu}$, $c\mu\bar{\nu}$, $c\tau\bar{\nu}$, proportional to V_{cb}; then the λ suppressed $b \to c\bar{u}s$,

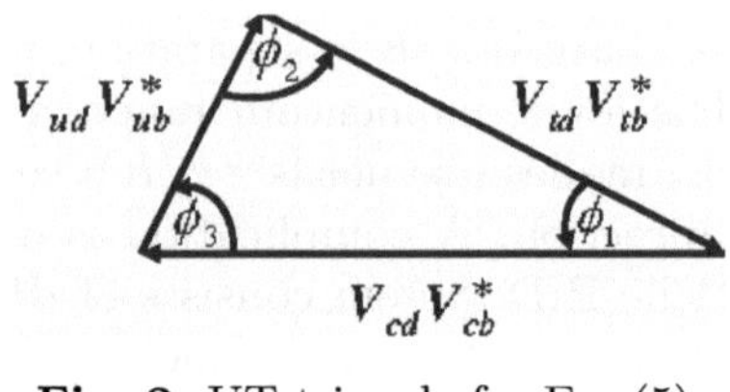

Fig. 2. UT triangle for Eq. (5).

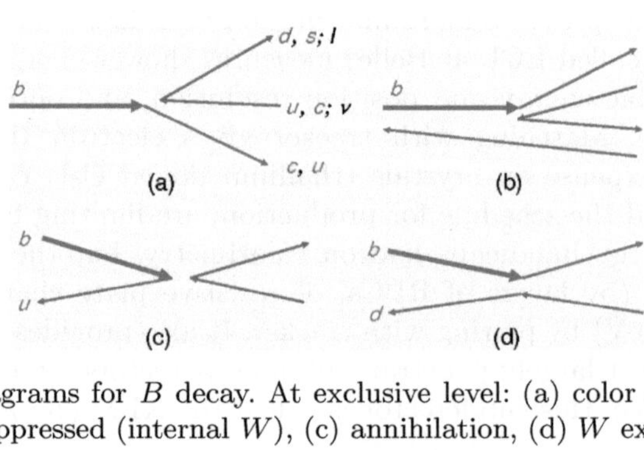

Fig. 3. Tree diagrams for B decay. At exclusive level: (a) color allowed (external W), (b) color suppressed (internal W), (c) annihilation, (d) W exchange.

$c\bar{c}d$; the V_{ub} suppressed $b \to u\bar{u}d$, $u\bar{c}s$, and semileptonic $b \to ue\bar{\nu}$, $u\mu\bar{\nu}$, $u\tau\bar{\nu}$. Finally, we have the λV_{ub} suppressed $b \to u\bar{u}s$, $u\bar{c}d$. The V_{ub} related hadronic modes can probe the CPV phase via tree decays.

At the exclusive level, e.g. two body final states, we illustrate further in Fig. 3 how final state mesons may form. The color suppressed mode has color mismatch between, say the quark from b decay and the spectator antiquark.

2.2 CPV Primer: Direct CPV

CPV is defined as a difference in probability between a particle process from the antiparticle process, e.g. between $B \to f$ and $\overline{B} \to \bar{f}$. It requires the presence of two interfering amplitudes, but besides the usual i from quantum time evolution, it needs complex dynamics. That is, the interference involves the presence of two kinds of phases. Let us illustrate how CPV occurs.

Consider the particle amplitude $A = A_1 + A_2$, where A_j have a CP invariant phase δ_j (i from QM) and a CPV phase ϕ_j (i from CPV dynamics). Absorbing an overall phase by defining $A_1 = a_1$ to be real, one has

$$\begin{aligned} A &= A_1 + A_2 = a_1 + a_2 e^{i\delta} e^{+i\phi}, \\ \overline{A} &= \overline{A}_1 + \overline{A}_2 = a_1 + a_2 e^{i\delta} e^{-i\phi}. \end{aligned} \tag{6}$$

The δ and ϕ are called the "strong" and weak phase, respectively. The QM or strong phase δ does not distinguish between particle or antiparticle, hence sign is unchanged. However, the dynamical or weak phase ϕ changes sign for the antiparticle process $\overline{A}$. This enrichment of quantum interference leads to an asymmetry between particle and antiparticle probabilities,

$$\mathcal{A}_{\rm CP} \equiv \frac{\Gamma_{\bar{B}^0 \to \bar{f}} - \Gamma_{B^0 \to f}}{\Gamma_{\bar{B}^0 \to \bar{f}} + \Gamma_{B^0 \to f}} = \frac{2a_1 a_2 \sin\delta \sin\phi}{a_1^2 + a_2^2 + 2a_1 a_2 \cos\delta\cos\phi}, \tag{7}$$

defined with respect to quarks. As $\mathcal{A}_{\rm CP}$ vanishes with either δ or $\phi \to 0$, CPV requires the presence of both CP conserving *and* CPV phases.

Eq. (6) is illustrated in Fig. 4, which shows geometrically how Eq. (7) materializes. If $\delta = 0$, then $A_1 + A_2$ and $\overline{A}_1 + \overline{A}_2$ are at angle ϕ above or below the real axis, hence of equal length. If $\phi = 0$, then $A_1 + A_2$ and $\overline{A}_1 + \overline{A}_2$

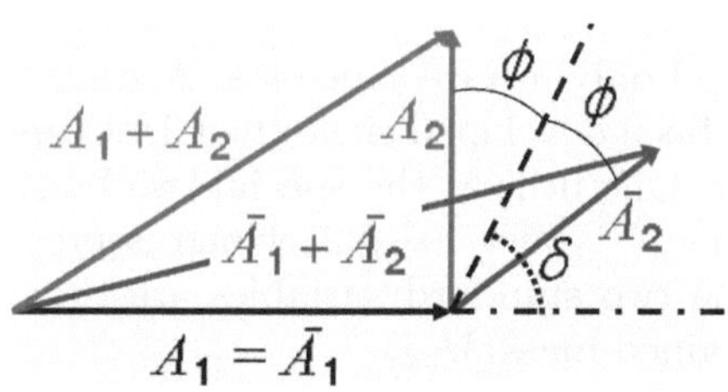

Fig. 4. Mechanism for CPV, Eqs. (6) and (7).

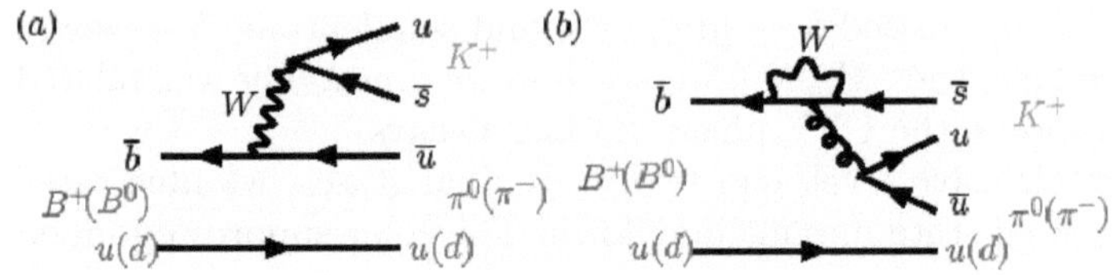

Fig. 5. Tree and Penguin $B \to K\pi$, $\pi\pi$ graphs.

are the same vector. Only when $\delta \neq 0$ *and* $\phi \neq 0$ does $|A_1 + A_2| \neq |\overline{A}_1 + \overline{A}_2|$ occur, which gives the asymmetry of Eq. (7).

In the KM model with 3 generations, the CPV phase is put in the 13 and 31 elements in the standard phase convention [6]. Thus, it is said that one needs the presence of all 3 generations to make CPV to occur.

Direct CPV in $B^0 \to K^+\pi^-$

Direct CPV (DCPV), which has recently been established in $B^0 \to K^+\pi^-$ (and $\pi^+\pi^-$ as well) mode, gives the most intuitive illustration. In Fig. 5 we show the tree (T) and penguin (P) diagrams for $B \to K\pi$, $\pi\pi$ decays. The tree $b \to u\bar{s}u$ diagram carries a weak phase via $\arg V_{us}V_{ub}^*$, while P is dominated by $V_{cs}V_{cb}^* \cong -V_{ts}V_{tb}^*$, which is almost real. If T and P have a relative strong δ phase, their interference would lead to *direct* (i.e., in decay amplitude itself) CPV. Indeed this was observed in 2004 [14].

The whimsical name “penguin” is attributed to a bet by John Ellis 30 years ago. If one complains that Fig. 5(b) bears no resemblance to a “penguin”, then neither does a Feynman diagram bear any resemblance to Feynman (although Feynman did pen it). We have already mentioned the discovery of the $B \to K^*\gamma$, $X_s\gamma$ penguins by CLEO II in 1993. The exclusive strong penguin modes $B \to K\pi$ [15], $\eta' K$ [16] were also discovered by CLEO II around 1998, and the strategy was clear that one searches for DCPV in rare decays. It took only 6 more years for DCPV in $B^0 \to K^+\pi^-$ to be measured between Belle and BaBar [14]. The electroweak penguin process $B \to K^{(*)}\ell^+\ell^-$, $X_s\ell^+\ell^-$, dominated by the Z penguin [17], were discovered [18] by Belle in 2002.

Rare B Reconstruction

DCPV can be observed only in rare processes. A standard procedure for rare B reconstruction has been developed since the CLEO era for two body decays such as $B \to K\pi$, $\pi\pi$. One utilizes the special two body $\Upsilon(4S) \to B\overline{B}$ decay kinematics: the B energy is just $E_{\rm CM}/2$, which is precisely known from the accelerator. Thus, the two standard variables are the energy difference ΔE and the beam constrained mass $M_{\rm bc}$,

$$\Delta E = \sum E_i - E_{\rm CM}/2, \qquad M_{\rm bc} = \sqrt{(E_{\rm CM}/2)^2 - \sum (\boldsymbol{p}_i)^2}, \tag{8}$$

where E_i and $\boldsymbol{p}_i$ are measured for particle i. A correctly reconstructed B meson event would peak in ΔE and $M_{\rm bc}$, as can be visualized by 1D projection plots, while background events would not. Note that the K and π in $B \to K\pi$, $\pi\pi$ are rather highly boosted, hence PID (ACC at Belle and DIRC at BaBar) performance is rather critical to separate the $K\pi$, $\pi\pi$ events.

With these standard techniques, it was a matter of time and providence for one to eventually catch the first DCPV measurement, which happened to be the $B^0 \to K^+\pi^-$ mode. The summer 2006 combined result gives [19]

$$\mathcal{A}_{K^+\pi^-} = -0.093 \pm 0.015, \text{ (Summer 2006)} \tag{9}$$

which is rather sizable, indicating a sizable CP invariant phase δ.

After CPV was discovered in K^0 mixing, it took 35 years for DCPV to be measured in kaon decay. For measurement of DCPV in B decay, it took only 3 year after the measurement of CPV in B^0 mixing in 2001.

2.3 Raison d'Être: $\sin 2\phi_1/\beta$ from $B^0 \to J/\psi K_S$

The billion dollar question for the B factories is to measure CPV phase ϕ_1/β via $B^0 \to J/\psi K_S$, which was fulfilled just 2 years after turning on!

We now need to explain the more intricate — and beautiful — mixing-decay interference [20]. We will identify what are δ and ϕ in Eqs. (6) and (7), the understanding of which should bring joy to the student.

Mixing-decay Interference

The final state $J/\psi K_S$ is a CP eigenstate. For an initial $\bar{B}^0$ meson, there are two paths for it to decay to a CP eigenstate $f_{\rm CP}$: direct $\bar{B}^0 \to J/\psi K_S$ decay (amplitude A), or via two steps, first $\bar{B}^0 \to B^0$ through mixing, then $\bar{B}^0 \to J/\psi K_S$ decay. In other words, denoting, the two paths $\bar{B}^0 \xrightarrow{A} J/\psi K_S$ and $\bar{B}^0 \xrightarrow{M_{12}} B^0 \xrightarrow{\overline{A}} J/\psi K_S$, where M_{12} is the $\bar{B}^0 \to B^0$ mixing amplitude, constitute the two interfering amplitudes A_1 and A_2 in Eq. (6).

From this we see that the CP invariant phase is from $\bar{B}^0 \leftrightarrow B^0$ oscillations, $e^{i\Delta mt}$, which is indeed quantum mechanical. The CPV weak phase comes from the dominant top contribution in the $b\bar{d} \to \bar{b}d$ box diagram, $M_{12} \propto V_{td}^{*\,2}$. The beauty of $\bar{B}^0 \to J/\psi K_S$ decay is that, the dominant tree level $b \to c\bar{c}s$ amplitude is proportional to $V_{cs}^* V_{cb}$ and practically real. This is by convention of Eq. (4) and hard to be affected by New Physics. Even for the penguin analogue of Fig. 5(b), the CKM factor is $V_{ts}^* V_{tb} \cong -V_{cs}^* V_{cb}$, which is also practically real in SM. Therefore, using Belle notation, the weak phase ϕ is effectively $\arg V_{td}^{*\,2} = \sin 2\phi_1$ to very good accuracy within SM. This is the reason why the measurement of time-dependent (mixing or oscillation dependent) CPV, or TCPV, became the *raison d'être* for the asymmetric B factory, and why good vertexing was placed at a premium.

Without further ado, we give the refinement of Eq. (7) for TCPV study in B^0 decays to CP eigenstate f [6]:

$$A(\Delta t) \equiv \frac{\Gamma_{\bar{B}^0(\Delta t)\to f} - \Gamma_{B^0(\Delta t)\to f}}{\Gamma_{\bar{B}^0(\Delta t)\to f} + \Gamma_{B^0(\Delta t)\to f}} = -\xi_f(\mathcal{S}_f \sin \Delta m\Delta t + \mathcal{A}_f \cos \Delta m\Delta t), \quad (10)$$

where ξ_f is the CP eigenvalue of f, $\Delta m = \Delta m_{B_d}$, $\bar{B}^0(\Delta t)$ is the state at time Δt for $\bar{B}^0(0) = \bar{B}^0$, and $\Delta t = t_{\rm CP} - t_{\rm tag}$ (more detail later), the time difference between "CP" side (e.g. $J/\psi K_S$) and flavor tagged side. The coefficient $\mathcal{S}_f$ associated with $\sin \Delta m\Delta t$ corresponds to $\sin\phi$ in Eq. (7), and $\mathcal{A}_f$ ($-\mathcal{C}_f$ for BaBar) associated with $\cos \Delta m\Delta t$ oscillation is related to DCPV, which could occur even at $\Delta t = 0$ (i.e. without oscillations). They are given by

$$\mathcal{S}_f = \frac{2\,\mathrm{Im}\,\lambda_f}{|\lambda_f|^2+1}, \quad \mathcal{A}_f = \frac{|\lambda_f|^2-1}{|\lambda_f|^2+1}, \quad \text{where } \lambda_f \equiv \frac{q}{p}\frac{\langle f|S|\bar{B}^0\rangle}{\langle f|S|B^0\rangle}. \quad (11)$$

Clearly one has $\mathcal{A}_f = \mathcal{A}_{\rm CP}$. Since $\langle f|S|\bar{B}^0\rangle$ is real for $f = J/\psi K_S$, one gets $\lambda_f \cong e^{-2i\phi_1}$, so *TCPV in* $\bar{B}^0 \to J/\psi K_S$ *is free of hadronic uncertainties.*

Measurement of TCPV in $b \to c\bar{c}s$

The $\Upsilon(4S)$ is only about 20 MeV above the $B\overline{B}$ threshold, so the transverse motions of the B's are insignificant. The vertex detector measures both decay vertices along the z axis, defined as the high momentum beam direction, and $\Delta t \cong \Delta z/\beta\gamma c$. The "$CP$ side" is reconstructed in, say, $J/\psi K_S$, the cleanest mode both experimentally and theoretically. The flavor has to be inferred from the "tag side", where one uses coherence to assign opposite flavor of tag side at $\Delta t = t_{\rm CP} - t_{\rm tag} = 0$. One then uses the time evolution of Eq. (10) to fit for $\mathcal{S}_f$ and $\mathcal{A}_f$. To summarize, one needs to

1) Reconstruct one B in CP eigenstate,
2) Tag the other B flavor,
3) Measure both decay vertices.

The effort is to reconstruct as many $b \to (c\bar{c})_{\rm charmonium}s$ final states with definite CP, including angular resolved $B \to J/\psi K^*$ to disentangle CP components. Correcting for ξ_f, one enlarges statistics to measure the so-called $\mathcal{S}_{c\bar{c}s}$, assuming that hadronic and other corrections are small.

$B \to J/\psi K_S \to \ell^+\ell^-\pi^+\pi^-$ final state is the easiest and cleanest to reconstruct because of the very narrow J/ψ width. Next, $B \to J/\psi K_L$ is reconstructed by associating activities in the KLM detector (IFR at BaBar) with no tracks (and potentially some activity in ECL), but the S/B gets worse. Correcting for the ξ_f factor, the combined result is called $\mathcal{S}_{J/\psi K^0}$. The collected events with a dataset of 535M $B\overline{B}$ pairs by Belle is shown in Fig. 6, where $B \to J/\psi K_S$ is evidently much cleaner. One also reconstructs in $\psi' K_S$, $\eta_c K_S$, $\chi_{c1} K_S$ as well as $J/\psi K^{*0}$ (with flavor nonspecific K^{*0} decays).

For flavor tagging, semileptonic decay provides the cleanest tag, but other tags, such as kaon (even Λ) flavor from the $b \to c \to s$ decay chain are

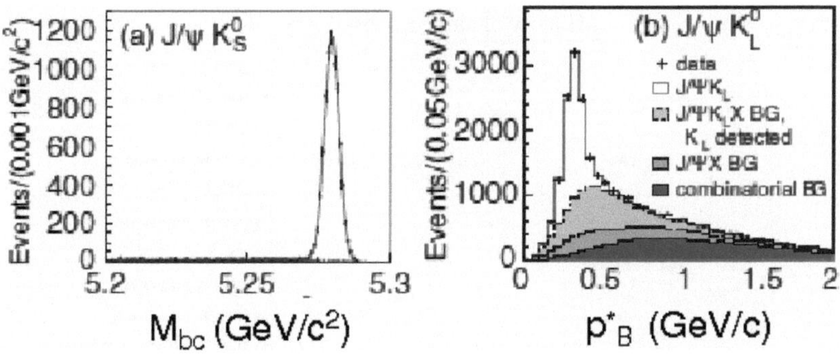

Fig. 6. (a) $M_{\rm bc}$ distribution of $B^0 \to J/\psi K_S^0$ candidate events in ΔE signal region, and (b) p_B^* (CM) distribution for $B^0 \to J/\psi K_L^0$ candidate events.

used, which has higher efficiency than lepton tags. The mistag probability w is defined as 1 minus the correctness of tag, which has to be measured, leading to a dilution factor $D = 1 - 2w$. At Belle one produces the $q = +/-$ assignment ($B^0/\bar{B}^0$ on tag side), with a purity of tag $r \in (0,\ 1)$.

Blind Analysis is worth special emphasis for B factories (and for all precision measurements, especially those with sign ambiguities). That is, the analysis is done "blind" to avoid possible bias. The to-be measured quantity is shifted *randomly* by some amount (within range of a few expected errors). All systematic studies are completed before "unblinding" for the final measurement.

The measurement of $\sin 2\phi_1/\beta$ ($\equiv \mathcal{S}_{c\bar{c}s}$ by the assumption that corrections are small) is as fundamental as the fine structure constant α. The world history of $\sin 2\phi_1/\beta$ measurement for 1998–2005 is shown in Fig. 7. The *raison d'être* for the B factories was achieved in year 2001.

To get a feel of data quality and current status, we plot in Fig. 8 the result for $B^0 \to J/\psi K^0$ with 535M $B\overline{B}$ pairs collected by Belle. Note the consistency between $B^0 \to J/\psi K_S^0$ and $J/\psi K_L^0$. The measured value is $\sin 2\phi_1 = 0.642 \pm 0.034 \pm 0.017$ for Belle, which is only for $B^0 \to J/\psi K^0$. BaBar continued to combine all $b \to c\bar{c}s$ modes, and the 2006 result is $\sin 2\beta = 0.714 \pm 0.032 \pm 0.018$, combining with Belle to 0.678 ± 0.026. Combining further with all measurements, the current world average is [19]

$$\sin 2\phi_1/\beta = 0.678 \pm 0.025, \quad \text{(Winter 2007)} \tag{12}$$

which now has a precision of 4%. This gives

$$\phi_1/\beta = (21.3 \pm 1.0)^\circ, \quad \text{(Winter 2007)} \tag{13}$$

up to sign ambiguities (68.7° is in principle allowed). Hence, $\sin 2\phi_1/\beta$ has now been precisely measured, with no sign of deviation (?) from SM, but note that there is some variance between the Belle and BaBar results.

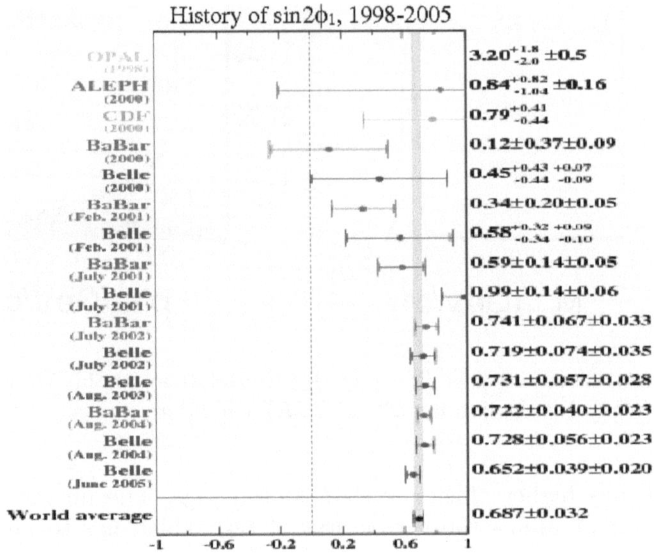

Fig. 7. World history of $\sin 2\phi_1/\beta$ measurement.

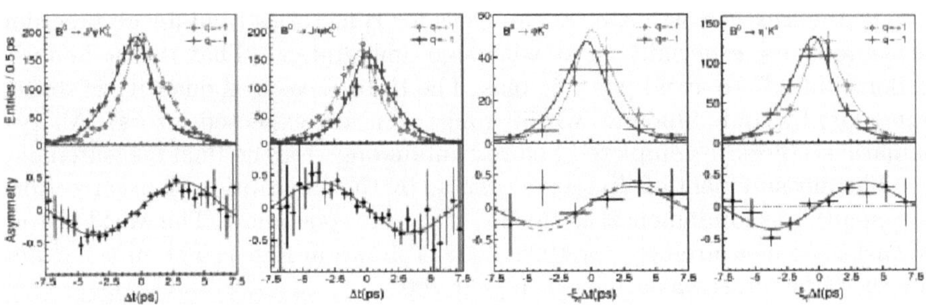

Fig. 8. Background-subtracted Δt distributions and asymmetries for (a), (b) $B^0 \to J/\psi K^0$ events of Fig. 6, and (c) ϕK^0, (d) $\eta' K^0$, with good tags ($r > 0.5$).

The question mark above refers to the tension between Eq. (12) and e.g. the indirect "measurement" of $\sin 2\phi_1/\beta = 0.752^{+0.057}_{-0.035}$ (Spring 2006) by fitting [21] to all data without the direct measurement input, which heightens if one takes the Belle result. Taken seriously, this could indicate New Physics.

2.4 $\sin 2\phi_1/\beta|_{\rm eff.}$ in $b \to s\bar{q}q$

Possible hints for New Physics emerged in the $\mathcal{S}_f$ parameter measurement in $b \to s\bar{q}q$ modes, which are dominated by the strong penguin P.

The excitement began with the summer 2003 result from Belle, based on 140 fb^{-1} data collected with the SVD1 vertex detector. With 68 ± 11 reconstructed $B \to \phi K_S$ events, Belle measured $\mathcal{S}_{\phi K_S} = -0.96 \pm 0.50^{+0.09}_{-0.11}$, which is rather different from $\mathcal{S}_{c\bar{c}s}$ value of Eq. (12), or the world average of 2003. However, dominated by a diagram like Fig. 5(b) with $g^* \to u\bar{u}$ replaced by an $s\bar{s}$ pair, the penguin amplitude $\propto V_{ts}^* V_{tb}$ in SM is also practically real, hence $\lambda_{\phi K_S} \cong q/p$ just as for $B \to J/\psi K_S$, hence

$$\mathcal{S}_{\phi K_S} \cong \sin 2\phi_1/\beta, \quad \text{(SM)} \tag{14}$$

is predicted. The Belle result showed a large deviation, and even the sign was opposite. Though not yet statistically significant enough (and with BaBar at variance), it aroused great interests, and many speculations about whether NP FCNC and CPV loop effects could break the equality of Eq. (14).

The large deviation indicated by Belle 2003 data turned out to be a mirage. By 2004, with doubling of the data, the measured value for $\mathcal{S}_{\phi K_S}$ vanished. It so happened that Belle installed the upgraded SVD2 vertex detector in summer 2003, and the almost equal amount of SVD2 data had opposite sign with the earlier SVD1 data. This does not mean the earlier data was wrong, since the number for that dataset did not change upon reanalysis. *Statistical fluctuations do happen*, and the only antidote is to have more data!

By 2006, the Belle analysis included $K_S \to \pi^0\pi^0$ as well as K_L, including $\phi \to K_S K_L$ reconstruction. Out of 535M $B\overline{B}$ pairs, about 310 $B \to \phi K_S$ and 110 ϕK_L events were reconstructed. The S/B is of course much poorer than Fig. 6. We give the Δt distribution and asymmetry in Fig. 8(c), which should be compared with Fig. 8(a). The measured result is

$$\mathcal{S}_{\phi K^0} = +0.50 \pm 0.21 \pm 0.06, \quad \text{(Belle 535M)} \tag{15}$$

which is 1σ below Eq. (12), but still consistent with Eq. (14). The BaBar 2005 result gives $+0.50 \pm 0.25^{+0.07}_{-0.04}$, but a different method was adopted in 2006. Fitting to low mass K^+K^- region for ϕK^0 and $f_0(980)K^0$ of three body $K^+K^-K^0$, gives $\mathcal{S}_{\phi K^0} = +0.12 \pm 0.31 \pm 0.10$, which is lower than Eq. (15).

Fig. 8(d) shows the $B^0 \to \eta' K^0$ result for Belle, giving

$$\mathcal{S}_{\eta' K^0} = +0.64 \pm 0.10 \pm 0.04, \quad \text{(Belle 535M)} \tag{16}$$

which, at 5.6σ this is the first observation of TCPV in a single charmless $b \to s$ mode. It is fully consistent with SM, i.e. the relation similar to Eq. (14). Unlike the large early fluctuation in ϕK^0 mode, here Belle has measured consistent results throughout. With 347M $B\overline{B}$ pairs, BaBar measures $\mathcal{S}_{\eta' K^0} = +0.55 \pm 0.11 \pm 0.02$, a 4.9σ effect, but it has migrated from a value consistent with zero in early phase, through the 2005 result of $+0.36 \pm 0.13 \pm 0.03$.

The $\Delta\mathcal{S}$ Problem

Many other $b \to s\bar{q}q$ modes are studied. No significant results for $\mathcal{A}_f$ $(= -\mathcal{C}_f)$ have been established so far. The emerging picture is a consistent systematic shift of $\mathcal{S}_f$ in these modes towards the lower side of $\mathcal{S}_{c\bar{c}s}$. In Fig. 9 we plot the

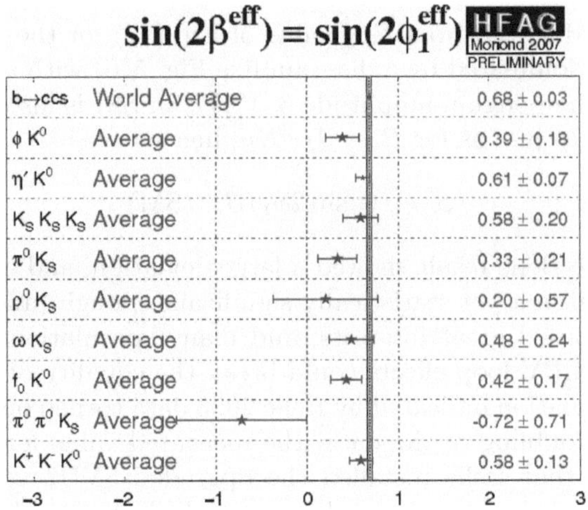

Fig. 9. HFAG Winter 2007 average result for $\mathcal{S}_{s\bar{q}q}$.

HFAG average results for $\mathcal{S}_f$ in various charmless $b \to s\bar{q}q$ modes, compared with $b \to c\bar{c}s$ (first row). All 9 charmless modes are lower than $\mathcal{S}_{c\bar{c}s}$, and echoes the original Belle splash of 2003. The naive average gives

$$\mathcal{S}_{s\bar{q}q} = +0.53 \pm 0.05, \quad \text{(Winter 2007)} \tag{17}$$

which can be compared with Eq. (12). In contrast, considering corrections to Eq. (14), theory expects a positive shift for SM. Given that the over -2σ difference between $\mathcal{S}_{s\bar{q}q}$ and $\mathcal{S}_{c\bar{c}s}$ has persisted for several years, and with Belle and BaBar generally in agreement, either there is some common experimental systematics, or it could be taken as a possible hint for New Physics. This has been one of the most important outcomes from B factories.

Again, more data would be needed, requiring something beyond LHCb, such as a Super B factory.

2.5 $\sin 2\phi_2/\alpha$ in $b \to u\bar{u}d$

A clean measurement for ϕ_2/α was originally projected, when the $b \to d\bar{u}u$ penguins, the analogues of $b \to s\bar{u}u$ were ignored. If so, then the formalism of Eqs. (10)–(11) would hold, just like for $b \to c\bar{c}s$ and $s\bar{q}q$, except that now one does have CPV phase in the decay amplitude, e.g. $\langle\pi^+\pi^-|S|\bar{B}^0\rangle/\langle\pi^+\pi^-|S|B^0\rangle \cong V_{ub}/V_{ub}^* = e^{-2i\phi_3}$. Combining with q/p, one gets $\lambda_{\pi^+\pi^-} = e^{+2i\phi_2}$, where $\phi_1 + \phi_2 + \phi_3 = \pi$ has been used.

$\pi\pi$ Program

Alas, the penguin cannot be ignored. Furthermore, there may be a strong phase δ between the tree and penguin amplitudes. The diagrams are just as in Fig. 5 except replacing $\bar{s}$ by $\bar{d}$. Retracing the above steps, one gets

$$\lambda_{\pi^+\pi^-} = e^{+2i\phi_2} \frac{T + Pe^{+i\phi_3} e^{i\delta}}{T + Pe^{-i\phi_3} e^{i\delta}}, \tag{18}$$

which collapses back to $e^{+2i\phi_2}$ as $P/T \to 0$. But it seems that $P/T \sim 0.3$, as can be seen from comparing the penguin dominant $B \to K\pi$ rate with the tree dominant $B \to \pi\pi$ rate, taking care of CKM factors. Furthermore, the presence of both weak and δ phases leads to DCPV, i.e. the prospect for a sizable $\mathcal{A}_f$ term in the TCPV fit. Thus, the $\pi^+\pi^-$ analysis would lead to only $\phi_2/\alpha|_{\text{eff.}}$ that is "polluted" by P/T in a complicated way.

Many improved methods were proposed to alleviate the problem. Eventually one needs to do an isospin, or SU(2), analysis by combining all information, both decay rate and CP asymmetries, from $\pi^+\pi^-$, $\pi^+\pi^0$ as well as $\pi^0\pi^0$ modes. This approach can remove the "penguin pollution" and disentangle ϕ_2/α from $\phi_2/\alpha|_{\text{eff.}}$. The discovery of sizable enhancement of the color-suppressed (see Fig. 3(b)) $\pi^0\pi^0$ mode gave hope to this program.

The measurement of ϕ_2/α, starting with the $\pi^+\pi^-$ mode, has a tortuous history to match theory development, and Belle and BaBar have never shown good agreement. The current Belle result is

$$\begin{aligned} \mathcal{S}_{\pi\pi} &= -0.61 \pm 0.10 \pm 0.04, \text{ (Belle 535M)} \\ \mathcal{A}_{\pi\pi} &= +0.55 \pm 0.08 \pm 0.05, \end{aligned} \tag{19}$$

based on ~1460 signal events, claiming 5.5σ and 5.6σ observations for DCPV and TCPV, respectively. BaBar gives

$$\begin{aligned} \mathcal{S}_{\pi\pi} &= -0.60 \pm 0.11 \pm 0.03, \text{ (BaBar 383M)} \\ \mathcal{A}_{\pi\pi} &= +0.21 \pm 0.09 \pm 0.02, \end{aligned} \tag{20}$$

based on ~675 $\pi^+\pi^-$ events, which give 5.5σ evidence for overall CPV, but only a mild hint for DCPV ($\mathcal{A}_{\pi\pi}$). We plot the history of $\pi^+\pi^-$ study in Fig. 10. It started with strong disagreement: Belle having a strong effect outside the physical boundary, while BaBar started consistent with zero. Now there is consensus that $\mathcal{S}_{\pi\pi} \sim -0.60$, but disagreement remains in $\mathcal{A}_{\pi\pi}$. It is said that, by SU(3) symmetry, the DCPV in $B^0 \to \pi^+\pi^-$ should be roughly 3 times as large as in $B^0 \to K^+\pi^-$. If such is the case, then the mean of the Belle and BaBar results seem correct.

The $\pi\pi$ program for ϕ_2/α measurement still lacks $\mathcal{S}_{\pi^0\pi^0}$ and $\mathcal{A}_{\pi^0\pi^0}$. No stringent constraint is obtained with $\pi\pi$ alone.

$\rho\pi$ Program

It had been claimed that an analysis in $B^0 \to \pi^+\pi^-\pi^0$ Dalitz plot, by interference between $\rho^\pm\rho^\mp$ and $\rho^0\pi^0$ modes, could give a clean measure of ϕ_2/α. The simplified early assumptions turned out also to be a false start.

In fact, the $\rho\pi$ program can be pursued, but the analysis is extremely complicated. We just mention that the latest study involves 26 fit parameters.

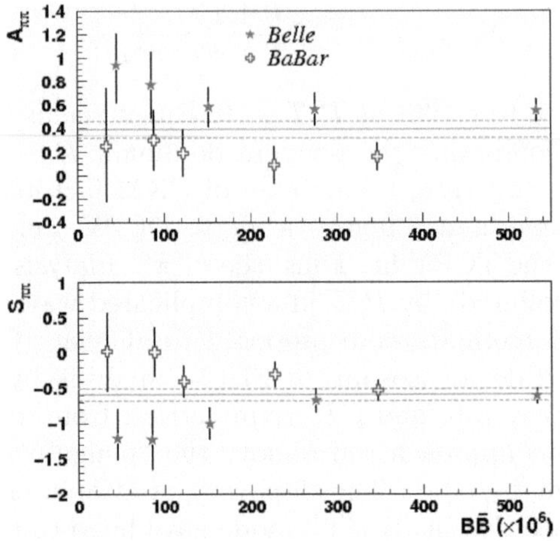

Fig. 10. History of TCPV study in $\pi^+\pi^-$ mode.

The Belle result based on 449M $B\overline{B}$ pairs with a Dalitz plot isospin analysis, give $67^\circ < \phi_2 < 95^\circ$, but other regions are allowed. The BaBar result, based on 375M $B\overline{B}$ pairs, gives $\alpha = (87^{+45}_{-13})^\circ$, with no constraint at 2σ.

$\rho\rho$ Program

With the difficulties in $\pi\pi$ and $\rho\pi$ programs, for some time the $\rho\rho$ program seemed the best ϕ_2/α program before 2006: $\mathcal{B}(B^0 \to \rho^+\rho^-)$ is large, almost 100% longitudinally polarized hence a pure CP eigenstate with no need for angular disentanglement. $\mathcal{B}(\rho^0\rho^0)$ seemed very small, hence no need for isospin analysis. Surprisingly, unlike $\rho\pi$ or modes involving kaons, there is little 3- or 4-body contamination by resonances. With such blessings, the summer 2004 and 2005 results, dominated by the BaBar study, were quite consistent with CKM fit results made without information of α in the fit.

Alas, things made the turn for the worse in 2006: BaBar found 3.5σ evidence for $\rho^0\rho^0$, and the situation became somewhat analogous to the $\pi\pi$ program, and the constraint on ϕ_2/α weakened to 92 ± 21 degrees, and $\rho\rho$ is no longer the "best program". One now needs information on $\mathcal{A}_{\rho^0\rho^0}$.

The combined result from $\pi\pi/\rho\pi/\rho\rho$ does give a useful result on ϕ_2/α. With the Belle $\rho\pi$ result yet to be included, one has

$$\phi_2/\alpha = (93^{+11}_{-9})^\circ, \tag{21}$$

which can be compared with the global fit result of $(98^{+5}_{-19})^\circ$ without information input from ϕ_2/α itself.

With all its complexity and uncertainties, the ϕ_2/α program is not a good place to discern New Physics effects. Discussion of measurement of the third phase ϕ_3/γ is deferred to Lecture 3.

2.6 New Physics in DCPV?

Regarding New Physics, we remark in passing that there may be a second hint for New Physics from the B factories, namely the aforementioned DCPV discussed in §**2.6**. Concurrent with the measurement of $\mathcal{A}_{K^+\pi^-}$, Eq. (9), it was found that the behavior for $B^+ \to K^+\pi^0$ was different

$$\mathcal{A}_{K^+\pi^0} = +0.047 \pm 0.026, \text{ (Summer 2006)} \tag{22}$$

which is consistent with zero, even opposite in sign to $\mathcal{A}_{K^+\pi^-}$. The difference

$$\Delta\mathcal{A}_{K\pi} \equiv \mathcal{A}_{K^+\pi^0} - \mathcal{A}_{K^+\pi^-} = 0.14 \pm 0.03 \tag{23}$$

is more than 5σ from zero. This certainly has something to do with enhancement of color-suppressed tree diagrams (Fig. 3(b)), as we have seen in $B^0 \to \pi^0\pi^0$. But it could also arise partially from New Physics through the electroweak penguin $P_{\rm EW}$ process that contributes to $B^+ \to K^+\pi^0$. The electroweak penguin is dominated by Z, or bsZ coupling. The virtual Z^0 can turn into a π^0 but not $\pi^\pm$, so differentiates between $\mathcal{A}_{K^+\pi^0}$ and $\mathcal{A}_{K^+\pi^-}$. Since the effective bsZ coupling within SM is real, to be able to generate Eq. (23) would require New Physics CPV!

3 A Broadband Pursuit of FPCP

Having covered the main part of CPV, in this section we cover many different topics to illustrate the richness of flavor physics explored at the B factories and beyond. Each subject can be touched upon only briefly.

3.1 Sides of Unitarity Triangles: V_{cb} and V_{ub}

The pattern $|V_{ub}|^2 \ll |V_{cb}|^2 \ll |V_{us}|^2$ lead to the Wolfenstein parametrization of Eq. (4). Measurement of V_{cb} is equivalent to measurement of A, while V_{ub} gives ρ and η, the latter can only be probed via CPV effects.

The precision of V_{cb} measurement was greatly improved by Heavy Quark Symmetry [22], which is rather analogous to heavy nucleus effect in atomic physics. The identical chemistry of isotopes corresponds to flavor symmetry, while the minuscule "hyperfine" splittings correspond to spin symmetry.

Imagine $B \to D^*\ell\nu$ decay. In the m_b, $m_c \to \infty$ limit, when a b quark decays to $c\ell\nu$, if the c quark remains stationary, while the ℓ and ν exit the meson practically without interacting, the "brown muck" or light degrees of freedom of the meson is unaffected. This zero-recoil limit corresponds to $w = 1$, where $2w\, m_B m_{D^*} = m_B^2 - m_{D^*}^2 - q^2$ with q^2 the lepton pair mass squared. In formula analogous to muon decay, one has a $B \to D^*$ "form factor" $F(w)$ that depends only on w in the heavy quark limit, and computable by lattice QCD. At the inclusive level, Luke's theorem says $1/m_b$ correction vanishes. These and other theoretical tools aid the precise determination of V_{cb}. But the actual discussion quickly gets technical. For all practical purposes (FAPP), $|V_{cb}| \simeq 0.04$, and $A \simeq 0.85$.

The measurement of $|V_{ub}|$ is much more difficult. The $b \to u\ell\nu$ rate is only at % level of $b \to c\ell\nu$, which becomes the main background. Proposed inclusive methods have been to impose cuts on the E_ℓ, m_X or q^2 spectrum above the m_D related threshold. Alternatively, one could measure exclusive $B \to (\pi,\ \rho,\ \omega)\ell\nu$ modes and face form factor uncertainties, with less aid from Heavy Quark Symmetry. At the moment, inclusive methods tend to give higher values than exclusive methods. FAPP for the nonspecialist, $\sqrt{\rho^2 + \eta^2} \sim 0.5$.

3.2 CPV Phase ϕ_3/γ: DK and $D\pi$ Methods

Many methods have been proposed for the measurement of the third CPV phase angle ϕ_3, or γ, of the unitarity triangle of Fig. 2 that corresponds to Eq. (5). But the most promising method in the long term, is the so-called "DK" method, based on interference between two tree diagrams.

Quantum interference between two amplitudes is always the key to the study of CPV, as illustrated in **§2.2** and Fig. 4. Here one has two *tree* amplitudes, so insensitive to New Physics effect. Consider $B^- \to D^{(*)0}K^{(*)-}$, $\bar{D}^{(*)0}K^{(*)-}$ decays. The first arises from CKM-favored $b \to c\bar{u}s$ (Fig. 3(a)) process $\propto V_{cb}V_{us}^* = A\lambda^3$, while the second arises from the CKM- *and* color-suppressed $b \to u\bar{c}s$ (Fig. 3(b)) process $\propto V_{ub}V_{cs}^* = A\lambda^3(\rho - i\eta)$. The two amplitudes interfere when $D^{(*)0}$ and $\bar{D}^{(*)0}$ decay to common final state f. Depending on f, three methods have been proposed to exploit this interference: GLW (D^0 decay to CP eigenstate), ADS (doubly Cabibbo suppressed D^0 decay) and "Dalitz plot analysis". We shall present only the latter [23], because the second amplitude is very suppressed hence requires good statistics. Furthermore, it is the most experimentally oriented.

Let us illustrate with $D^0 \to K_S\pi^+\pi^-$, which has a large branching fraction. $B^\mp \to [K_S\pi^+\pi^-]_D K^\mp$ decay can be described by the amplitude,

$$M_\mp \propto f(m_\mp^2, m_\pm^2) + r\, e^{i\delta} e^{\mp i\phi_3}\, f(m_\pm^2, m_\mp^2) \qquad (24)$$

where $A(B^- \to \bar{D}^0K^-)/A(B^- \to D^0K^-) \equiv r\, e^{i\delta}e^{-i\phi_3}$, and $f(m_-^2, m_+^2)$ describes the Dalitz plot amplitude for $D^0 \to K_S\pi^+\pi^-$, with $m_\pm^2 \equiv m^2(K_S\pi^\pm)$. The latter can be determined from flavor-tagged $D^{*+} \to D^0\pi^+$ decays, but requires some modelling. With very large data sample, a bin-by-bin analysis [23] can remove model dependence. We stress again that the method is powerful because r is very small, and the method includes GLW and ADS regions. Experimental studies pair up D, D^* and K, K^* modes.

An analogous approach is the $D\pi$ method for $2\phi_1 + \phi_3$. This process involves the allowed $b \to c\bar{u}d$ decay, interfering with a λV_{ub} suppressed $\bar{b} \to \bar{u}c\bar{d}$

decay brought about by B^0–$\bar{B}^0$ mixing, hence the extra $2\phi_1$ factor. Here r is extremely suppressed, hence the method is less fruitful at present.

The combined result on ϕ_3 gives

$$\phi_3/\gamma = (71^{+22}_{-30})^\circ, \tag{25}$$

which is the least certain among the three CPV phases. Adding up Eqs. (13), (21) and (25), one has

$$\phi_1 + \phi_2 + \phi_3 \sim 185^\circ, \tag{26}$$

which is consistent with closing, but still with large errors. One would need LHCb and likely a Super B factory to reach convincing precision that Eq. (5) indeed closes as a triangle.

3.3 Measurement of Large B_s Mixing (vs $b \to d$ Penguins)

Ever since the observation of large B^0 mixing, it was clear that B_s–$\overline{B}_s$ oscillations would be very rapid hence hard to measure. There is at least the enhancement factor of $|V_{ts}/V_{td}|^2 \sim \lambda^{-2} \sim 21$. At the end of the LEP era (the rapid oscillations make measurement at B factories infeasible), the lower bound for Δm_{B_s} was roughly at 14 ps^{-1}, or about 18 times Δm_{B_d}.

The Tevatron Run II should have the capability to probe the SM range of 15 – 30 ps^{-1}. It was finally delivered in 2006. With the first hint from DØ, the CDF experiment reported evidence, then observation [24], of B_s mixing,

$$\Delta m_{B_s} = 17.77 \pm 0.10 \pm 0.07 \text{ ps}^{-1}, \tag{27}$$

which is already a precision measurement. The result, using the amplitude scan method (a filter), is shown in Fig. 11.

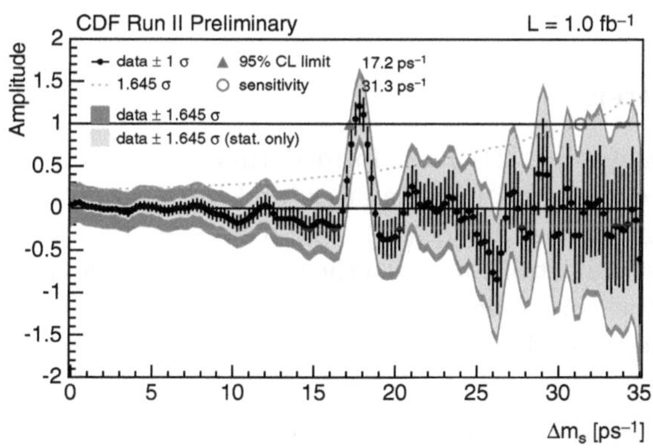

Fig. 11. Observation of large B_s mixing with amplitude scan method (from [24]).

Besides reconstruction in semileptonic modes like $B_s \to D_s \ell \nu$, CDF has the advantage of reconstruction in hadronic modes such as $D_s \pi$, which has better proper time resolution (25μm) than in semileptonic mode. Furthermore, besides opposite side tagging (similar to B factories), CDF enjoys same side tagging of the B_s flavor by the associated $K^\pm$, improving the overall effectiveness, εD^2, where ε is efficiency and $D = 1 - 2w$ is dilution.

The measured Δm_{B_s} is consistent with SM, but it can be said that it is slightly on the low side, when compared to predictions made beforehand. With the measured Δm_{B_d}, one can extract

$$\left| \frac{V_{td}}{V_{ts}} \right| = \xi \sqrt{\frac{\Delta m_{B_d}}{\Delta m_{B_s}} \frac{m_{B_s}}{m_{B_d}}} \simeq 0.21, \tag{28}$$

where $\xi = \sqrt{f_{B_s}^2 B_{B_s} / f_{B_d}^2 B_{B_d}}$ is a hadronic parameter ratio that has to be calculated, e.g., by lattice methods, which dominates the error on $|V_{td}/V_{ts}|$.

The question of interest is now the size of CPV in B_s system. Tevatron Run II would end by 2009, with up to 8 fb^{-1} expected per experiment, while LHCb will start running with LHC commissioning in 2008. LHCb should be able to repeat the "CDF measurement" of Δm_{B_s} in an "instant", and should be able to probe $\sin 2\phi_{B_s}$ at the SM level of -4%. However, if $|\sin 2\phi_{B_s}|$ is large, say 0.5–0.7, perhaps the Tevatron could draw another punch?

$b \to d\gamma$ and $B \to K\overline{K}$

Another probe of $|V_{td}/V_{ts}|$ is $b \to d\gamma$. $B \to K^*\gamma$ was already established in 1993, while $B \to \rho\gamma$ appeared at the B factories in 2006. At the moment there is still some conflict between Belle and BaBar on the actual strength. At the same time, the gluonic $b \to d\bar{s}s$ penguin processes (changing $\bar{s}$ to $\bar{d}$, and the $u\bar{u}$ pair from gluon to $s\bar{s}$) $B \to \bar{K}^0 K^0$, $\bar{K}^0 K^+$ have also been established by Belle and BaBar. However, the precision for $|V_{td}/V_{ts}|$ determination is not competitive with $\Delta m_{B_d}/\Delta m_{B_s}$.

The Two "Fitters"

Two competing groups [21] have formed that try to combine all data and all theory approaches and uncertainties, and display the information on the unitarity triangle of Fig. 2 (Eq. (5)). The results are displayed in Fig. 12. We shall not go into the details, nor into the debate of "frequentist" or "Bayesian" approach. At the core, one can see the unitarity triangle. At worst, these figures are some form of "Modern Art".

3.4 Large Missing Energy Decays: $B \to \tau\nu$ etc.

In discussing B_s mixing and V_{td}/V_{ts}, we mentioned hadronic parameter dependence such as on f_{B_q}, the meson decay constant. The $B^\pm \to \tau^\pm \nu$ decay,

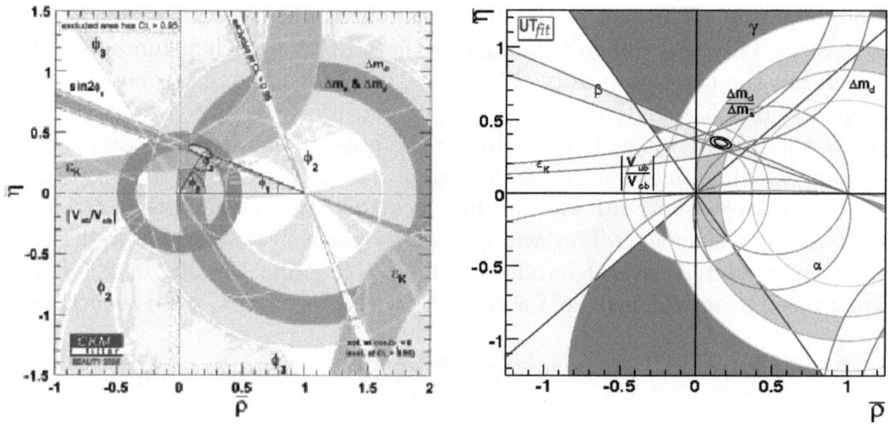

Fig. 12. The CKMfitter and UTfit representation of current information on the unitarity triangle of Eq. (5) (from relevant web pages [21]).

entirely analogous to $\pi^\pm \to \mu^\pm\nu$ and $K^\pm \to \mu^\pm\nu$, probes the B meson decay constant directly. The process goes through an annihilation diagram as in Fig. 3(c), where the b and $\bar{u}$ quarks annihilate into a W^- boson, which converts into $\tau^-\nu$. The annihilation probes the wave function at the origin, where the meson decay constant is a measure.

The formula for $B^\pm \to \tau^\pm\nu$ decay is easily related to muon decay, except one now has only two body phase space. We have

$$\Gamma(B \to \tau\nu) = \frac{G_F^2}{8\pi} m_B m_\tau^2 \left(1 - \frac{m_\tau^2}{m_B^2}\right) f_B^2 |V_{ub}|^2, \tag{29}$$

which depends only on the product $f_B|V_{ub}|$. All hadronic dependence is absorbed into f_B, which, if computed from lattice, then $|V_{ub}|$ can be inferred from the measured rate. The nominal SM expectation is around 10^{-4}. A measurement of $\mathcal{B}(B \to \tau\nu)$ also probes New Physics, especially the effect of a charged Higgs H^+ boson, by simply modifying Eq. (29) by a factor [25]

$$\mathcal{B}(B \to \tau\nu) = \mathcal{B}^{\rm SM}_{B\to\tau\nu} \times r_H, \qquad r_H = \left(1 - \frac{m_B^2}{m_{H^+}^2} \tan^2\beta\right)^2, \tag{30}$$

where $\tan\beta$ is the usual ratio of vacuum expectations in models with two Higgs doublets (2HDM), such as in supersymmetry. The enhancement factor appears because of coupling to m_b and m_τ in the SUSY type of 2HDM.

The challenge for measuring $B \to \tau\nu$ is as follows. Neutrino detection techniques have been developed when there is only one neutrino. Since τ decay always involve a tau neutrino, the event has at least two neutrinos, hence has both large missing energy and missing mass. However, since B mesons decay semileptonically for over 20% of the time, if one misses any particle (track) from the other B meson in the $B\overline{B}$ event, it would become background for $B \to \tau\nu$. This was difficult enough in the CLEO era when luminosity was lower. For the high luminosity B factories, detectors are not hermetic enough. Not only one has myriads of cracks, one can see from Fig. 1 that the final focus quadrupole magnets enter inside the guts of the detector, and there is a finite opening angle for active detection.

The price to be paid was the full reconstruction of the other B, at the cost of 10^{-3} efficiency. This was pursued vigorously once Belle and BaBar each accumulated several hundred million $B\overline{B}$ pairs. Finally, in 2006 Belle announced 3.5σ evidence [26] for $B \to \tau\nu$, with $\tau \to e, \mu, \pi, \rho$ plus neutrino(s),

$$\mathcal{B}(B \to \tau\nu) = (1.79^{+0.56+0.39}_{-0.49-0.46}) \times 10^{-4}, \qquad \text{(Belle 2006)} \tag{31}$$

with BaBar value lower and still lack of evidence [19]. The combined result is lower than Belle's first evidence, which is consistent with SM expectation. If one treats f_B and $|V_{ub}|$ as known, one gets a range for r_H, which constrains the m_{H^+}–$\tan\beta$ plane, and can be improved in the future.

Together with $b \to s\gamma$, B physics provides powerful constraints on New Physics such as H^+ bosons.

$B \to K^{(*)}\nu\nu$ or $K^{(*)}+$ Nothing

Full reconstruction of one B opens up the study of other interesting channels. The $B \to K^{(*)}\nu\nu$ decay is rather similar to $B \to \tau\nu \to \pi(\rho)\nu\nu$ in final state selection: $K(\pi)$ in place of $\pi(\pi)$. Thus, it has been searched for concurrently.

In SM, the $B \to K^{(*)}\nu\nu$ process is induced by [17] bsZ penguin and box diagrams, analogous to the $B \to K^{(*)}\ell^+\ell^-$ process (which has the additional $bs\gamma$ penguin contribution). The expectation is at the 10^{-5} level. However, the experimental signature is just $B \to K^{(*)}+$ *nothing*. This *nothing* could also be *light dark matter*. It is interesting to note that, although the DAMA hint for dark matter in our solar neighborhood seems ruled out by the CDMS experiment, direct dark matter searches are not sensitive to $M < 10$ GeV. Thus, $B \to K^{(*)}\nu\nu$ study provides a window on dark matter search.

The experimental technique is rather similar to $B \to \tau\nu$ study: full reconstruction of "tag side" and $K+$ nothing on the signal side, Belle has searched with 535M $B\overline{B}$ pairs and reported preliminary results. The upper limits reached is still 10 times the SM expectation. Put in different way, there is still room for discovery. But progress would require a Super B factory.

Also motivated by dark matter search, Belle has made specific runs on the $\Upsilon(3S)$ to search for $\Upsilon \to$ "*nothing*" decays, via $\Upsilon(3S) \to \pi^+\pi^-\Upsilon$ using $\pi^+\pi^-$ as tag. No signal was found [27], but it illustrates the prowess and versatility of B factories. The special data taking took just a few days.

3.5 Some Tau/Charm Topics

$\tau \to \ell\gamma$

Radiative $\tau \to \ell\gamma$ decays are of great interest because it is extremely suppressed by the tiny neutrino mass within SM context. Analogous to the $b \to s\gamma$ transition, in supersymmetric models, for example, sneutrino-chargino or slepton-neutralino loops could induce the transition, and whether $\tau \to e\gamma$ or $\tau \to \mu\gamma$ is more important is model dependent, the branching fraction typically scaling as $\tan^2\beta/M_{\rm SUSY}^4$. As B factories are also τ factories, current limits have reached the 10^{-8} level, constraining the $\tan\beta$–$M_{\rm SUSY}$ plane. This complements the $\mu \to e\gamma$ and muon $g-2$ studies.

Many other lepton number (even baryon number) violating studies have been pursued.

D^0–$\bar{D}^0$ Mixing

The K^0–$\bar{K}^0$ mixing phenomenon is repeated in the B_d and B_s systems. The only other similar meson is the D^0, composed of $c\bar{u}$. For this system, $x_D \equiv \Delta m_D/\Gamma_D$ is highly suppressed within SM. Unlike K and B mesons, Γ_D has no CKM suppression. On the other hand, Δm_D is highly suppressed by the GIM mechanism: the d and s quarks are almost degenerate on the weak scale, and their effects very effectively cancel between $V_{ud}V_{cd}^* \cong -V_{us}V_{cs}^* \cong -\lambda$, while for b quark in the box diagram, the CKM factor $|(V_{ub}V_{cb}^*)^2| < \lambda^{10}$ is highly suppressed. Thus, one expects $x_D^{\rm SM}$ to be very small, hence a good probe of New Physics if x_D is dominated by short distance effects. However, SU(3) violation in D^0 and $\bar{D}^0$ decay to 2- and 4-body common final states could generate $\Delta\Gamma_D$, or mixing in rate.

After having a hint for some time, both Belle and BaBar have now found evidence pointing to $y_D \equiv \Delta\Gamma_D/2\Gamma_D$. Belle finds evidence [28] for y_{CP} ($= y$ if CPV small), by comparing D^0 lifetimes in K^+K^-, $\pi^+\pi^-$ channels vs $K^+\pi^-$. BaBar finds evidence [29] for y' in $D^0 \to K^+\pi^-$ decay, but together with a negative x'^2 with a large correlation between the two. Taken together, these results imply that y_D seems to be at the % level.

Having $y_D \sim$ % level could generate x_D at similar level [30]. But this would be long distance, SU(3) breaking physics, and less interesting. The real interest is in x_D generated by short distance effect, i.e. from New Physics. It seems that, again, CPV would be the path to pursue, but this would require much higher luminosities as there is no indication of it so far.

$D_q^+ \to \ell^+\nu$ and f_{D_q}

Whether short or long distance effect, the D meson decay constant f_D enters decay and mixing amplitudes. Furthermore, comparing between experiment and lattice results for f_{D_q} provides an important crosscheck of lattice calculations for f_B, which is important for extracting information from B_d and B_s mixing, as well as $B \to \tau\nu$. Thus, measurement of f_{D_q} is of great interest.

The rate $\Gamma(D_q^+ \to \ell^+\nu)$ is of the form of Eq. (29) by trivial replacement. Unfortunately, because there is no $\tan\beta$ enhancement of the charged Higgs coupling to charm, while the d and s quark masses are too light, there is no sensitivity to H^+ like in $B \to \tau\nu$. The current experimental value for f_{D_s} is of order 280 MeV, while f_{D^+} is of order 220 MeV, which are higher than lattice results, but the ratio $f_{D_s}/f_{D^+} \sim 1.26$ is in better agreement with lattice.

Spectroscopy — Spectacular Renaissance!

The discovery of new charmed-strange (D_{sJ} [3]) and charmonium ($X(3872)$ [2]) mesons have lead to a renaissance, after a 25 year hiatus, in meson spectroscopy, which we cannot do justice here, because of space, and because we are concerned with flavor rather than aspects of strong interactions.

There are now new D_{sJ} mesons decaying to DK, new baryons decaying to Dp, new charmed baryons decaying to $\Lambda_c K\pi$, etc., and there is no good interpretations yet. On the charmonium side, besides the narrow $X(3872)$ found by Belle, a broad 1^{--} $Y(4260)$ state was found by BaBar. Both states decay to $\pi^+\pi^- J/\psi$. A proliferation of states has followed: $X(3940)$, $Y(3940)$, $Y(4350)$, etc. While it is not yet clear how to place these states, the future is clearly wide open.

4 Future Perspective

Flavor physics is reaching a new juncture, and will undergo a major change after 2008. Some facilities would be closing down, some would experience rebirth or transformation. And, there will be the new kid in town: LHCb. One should also take the broader view and, together with the other two subjects of electroweak symmetry breaking and mass generation, neutrino physics and lepton sector, see flavor physics within the common quest to unravel New Physics.

Closings

With the Linac used as part of the new LCLS light source, PEP-II and BaBar will stop in 2008. The CLEO-c detector, morphed from CLEO-III by pressure of the B factories, would also end after taking some $\psi(3770) \to D\overline{D}$ data.

With LHC physics run to start in 2008, the Tevatron would also close down around that time. But Tevatron B physics would continue to deliver good results in $\mathcal{A}_{K^-\pi^+}$, could pull a coup in CPV in $B_s \to J/\psi\phi$, and could make other discoveries such as DCPV in $B^- \to J/\psi K^-$, as well as "owning" heavy b hadrons until LHC turn-on.

Rebirth

With CLEO-c moving into its ground, BEPC and BES was pushed to upgrade. With two rings, BEPC-II is designed to reach 10 times the luminosity at CLEO-c, and commissioning will start in 2007. The detector also undergoes a major upgrade to BES-III, aimed for the physics of charm and charm spectroscopy starting in 2008.

As the brightest collider in the world, and with superb physics output, KEKB and Belle aim for a well-deserved major upgrade to a Super B factory, with increase of luminosity by two orders of magnitude. A plan ("Grand Lepton Collider Project") has been worked out with KEK to integrate and optimize accelerator manpower usage between SuperKEKB and ILC (as well as J-PARC), where work on SuperKEKB can be viewed as an early phase of ILC accelerator manpower buildup. The problem is a budget shortfall in maintaining such an ambitious program, especially during 2008–2010 when KEK also has to run the J-PARC proton facility which is just being finished. Hopefully, with the success of the Japanese particle physics program, and the overall growth of Asia, the ambitious program can be carried through.

While one can be optimistic for SuperKEKB, there is another "Linear Collider"-inspired SuperB proposal at INFN, Italy. With low emittance, low beta and low current, the novel design is quite different from the more traditional high current, high beam-beam parameter SuperKEKB.

Newborn

With "regeneration" pains on the e^+e^- front, hadronic collider B facilities are entering a new phase: LHCb. With the capabilities of B studies in a hadronic environment demonstrated at the Tevatron, LHCb capitalizes on the high b cross section and optimize to a forward detector design, as well as choice of luminosity. It has efficient trigger for hadronic and leptonic modes, excellent tracking and vertexing, and two RICH (Ring Imaging Cherenkov) detectors for excellent PID. What LHCb is somewhat weaker on is photons, or ECAL.

LHCb vs Super B

So, the future of flavor physics seems to be LHCb vs a Super B factory. In fact, they are quite complementary.

With much larger production cross section and the large boost, together with triggering, vertexing and PID capabilities, LHCb can do very well in most processes not involving γ (hence π^0) or neutrinos, and if there are not too many tracks. In particular, important topics of B_s system, such as TCPV, $B_s \to \mu^+\mu^-$, and especially heavy b hadron spectroscopy, would be the domain of LHCb. Although B factories can run on the $\Upsilon(5S)$ to study B_s physics, it would not be easy to compete with LHCb and exclusively run at that energy. On traditional programs such as ϕ_1/β, ϕ_2/α, ϕ_3/γ for B_d and B_u systems, and DCPV in inclusive $b \to s\gamma$, the two approaches are complementary. When there is more than one photon, or on subtleties such as the lack of knowledge of the B direction in hadronic production, Super B would prevail. Examples are $\mathcal{S}_{\eta' K_S}$, and especially $\mathcal{S}_{K_S \pi^0 \gamma}$, where Super B has the advantage of knowing the B direction (basically the boost direction) for forming the "K_S" vertex. When it comes to processes involving neutrinos, such as $B \to K^{(*)}\nu\nu$, $\tau\nu$, $D^{(*)}\tau\nu$ etc., then LHCb is powerless. This also holds for inclusive processes, where the clean e^+e^- environment and the exclusive $B\overline{B}$ final state, are great advantages. Similarly, the clean environment gives Super B an edge on tau/charm topics such as $\tau \to \mu\gamma$ and D^0–$\bar{D}^0$ mixing.

An observation illustrates the point: the first few years of running of LHCb would not drastically improve the significance of the $\Delta\mathcal{S}$ problem. LHCb could approach $\mathcal{S}_{\phi K_S}$, but not very well, because of the relatively long-lived nature of $K_S \to \pi^+\pi^-$, while $\phi \to K^+K^-$ is too close to threshold, so the two kaon tracks do not separate well at production. For $K_S\pi^0$, again one cannot find a vertex, while for $\eta' K_S$, reconstruction of η' is not so easy, and one still has the K_S to deal with. Thus, improved understanding of whether there is a genuine problem in $\Delta\mathcal{S}$ requires a Super B factory. It remains to be seen how well can LHCb do with DCPV involving π^0s in final state, such as $\mathcal{A}_{B^+ \to K^+\pi^0}$. For these issues, and other more practical ones, the LHCb experiment is already discussing upgrade possibilities to improve detector performance.

To say the least, the superb hadronic collider approach of LHCb would greatly enrich flavor studies, especially in B_s system. But a Super B factory would complement the program and make it more complete. This would remain true for the next decade or two.

The Broad Perspective

The flavor program should be viewed in the broader perspective of our search for physics beyond SM, i.e. New Physics.

The high energy frontier will open up greatly with the LHC, which has as its main goal the search for the Higgs boson, to understand electroweak symmetry breaking and the origins of mass, where SUSY or other phenomena such as large extra dimensions may be uncovered. If LHC uncovers new particles, the study of virtual effects, and the effects on B, D, τ flavor, as well as kaon and muon sectors, would be of common interest.

It can be said that New Physics has already been seen in the neutrino sector: neutrinos have mass [1]. Thus, the neutrino sector itself has a CKM-like mixing and CPV program to be explored, and one current focus is to try to measure $\sin 2\theta_{13}$, since the nonvanishing of which is needed for the continued CPV search. Neutrino mass also touches on many subjects, from cosmology, to lepton flavor violation. These become common cause not only with the Super B factory program of lepton flavor and CP violation in τ sector, such as $\tau \to \ell\gamma$, but also the $\mu \to e\gamma$ and muon $g-2$ programs.

5 Conclusion

We have covered the current activities on the flavor and B physics front, from the Standard Model predictions, to potential New Physics indications. The B factories have performed spectacularly, as we have covered in the main part of CPV studies of ϕ_1, ϕ_2 and ϕ_3. The main goal of the B factories, the measurement of $\sin 2\phi_1$, has already become a precision measurement!

All of us are of course more interested in uncovering New Physics, as implied by the title of this School. The detailed study of SM predictions, such as overconstraining the CKM unitarity triangle, is a more conservative and painstaking approach. The other, more frontal kind of approach is illustrated by the $\Delta\mathcal{S}$ problem, where experimental measurement suggests that it may be negative, while theoretical modelling suggests the opposite. Alas, the hint is only at 2.6σ, and cannot be drastically improved for the next few years.

We have advocated a second possible hint for New Physics, in the difference of $\Delta\mathcal{A}_{K\pi} = \mathcal{A}_{B^+\to K^+\pi^0} - \mathcal{A}_{B^0\to K^+\pi^-} \neq 0$. This can indeed be accounted for partially with an enhanced color-suppressed amplitude C, but the enhancement needed would be very great indeed, especially if one also wishes to touch the $\Delta\mathcal{S}$ problem. In fact $\Delta\mathcal{A}_{K\pi}$ is a window on New Physics from the electroweak or Z penguin: the virtual Z converts to π^0 but not π^-, hence affecting $\Delta\mathcal{A}_{K\pi}$, and probes New Physics CPV in the bsZ coupling. Since the box diagram and the Z penguin are closely related, from this perspective I expect interesting TCPV effects could emerge in B_s mixing as well as D^0 mixing, and maybe in DCPV in $B^+ \to J/\psi K^+$.

It is hard to foresee where New Physics would first emerge. The LHC is a good bet, but the LHC could also just uncover SM Higgs (or not even that!), and we could be left with flavor violation studies as our best probe for the New Physics landscape. Alternatively, it is not impossible that the Tevatron could uncover evidence for large TCPV in B_s system, such as $|\sin 2\Phi_{B_s}| \sim 0.5$ or higher, before LHC turn on. New Physics involving flavor would come to the forefront, and the energy frontier opened by LHC would explore it further.

We are entering a potentially very exciting era.

Acknowledgements

I sincerely wish to thank Meltem Serin-Zeyrek, as well as the co-organizers and all the supporting workers of ISSMB06, which made the meeting at Akyaka such an enjoyable one. My only regret is that I did not have time to stay long enough, hence I look forward to a more leisurely return in the future. Hopefully the School becomes a regular one. I have enjoyed Belle very much as a collaboration, where some of the figures have been used freely. This work is supported by the National Science Council and the Ministry of Education of Taiwan.

References

1. H.T. Wong, this volume.
2. S.K. Choi *et al.* [Belle Collab.], Phys. Rev. Lett. **91**, 262001 (2003).
3. B. Aubert *et al.* [BaBar Collab.], Phys. Rev. Lett. **90**, 242001 (2003).
4. E. Fernandez *et al.* [MAC Collab.], Phys. Rev. Lett. **51**, 1022 (1983); N.S. Lockyer *et al.* [MARK II Collab.], *ibid.* **51**, 1316 (1983).
5. S. Stone, plenary talk at Lepton-Photon Symposium, Cornell, USA, 1983.
6. W.M. Yao *et al.* [Particle Data Group], J. Phys. G **33**, 1 (2006).
7. H. Albrecht *et al.* [ARGUS Collab.], Phys. Lett. B **192**, 245 (1987).
8. R. Ammar *et al.* [CLEO Collab.], Phys. Rev. Lett. **71**, 674 (1993); M.S. Alam *et al.* [CLEO Collab.], *ibid.* **74**, 2885 (1995).
9. S. Bertolini, F. Borzumati and A. Masiero, Phys. Rev. Lett. **59**, 180 (1987); N.G. Deshpande *et al.*, *ibid.* **59**, 183 (1987).
10. B. Grinstein and M.B. Wise, Phys. Lett. B **201**, 274 (1988); W.S. Hou and R.S. Willey, *ibid.* B **202**, 591 (1988).
11. Attributed to P. Oddone, 1987.
12. F. Abe *et al.* [CDF Collab.], Phys. Rev. Lett. **68**, 3403 (1992).
13. A. Go *et al.* [Belle Collab.], quant-ph/0702267.
14. B. Aubert *et al.* [BaBar Collab.], Phys. Rev. Lett. **93**, 131801 (2004); Y. Chao *et al.* [Belle Collab.], *ibid.* **93**, 191802 (2004).
15. R. Godang *et al.* [CLEO Collab.], Phys. Rev. Lett. **80**, 3456 (1998).
16. B.H. Behrens *et al.* [CLEO Collab.], Phys. Rev. Lett. **80**, 3710 (1998); T.E. Browder *et al.* [CLEO Collab.], *ibid.* **81**, 1786 (1998).
17. W.S. Hou, R.S. Willey and A. Soni, Phys. Rev. Lett. **58**, 1608 (1987).
18. K. Abe *et al.* [Belle Collab.], Phys. Rev. Lett. **88**, 021801 (2002); J. Kaneko *et al.* [Belle Collab.], *ibid.* **90**, 021801 (2003).
19. The latest result, including many measurements of rare decays, can be found at the HFAG (Heavy Flavor Averaging Group) webpage.
20. A.B. Carter and A.I. Sanda, Phys. Rev. Lett. **45**, 952 (1980).
21. See the webpages of http://www.slac.stanford.edu/xorg/ckmfitter/, and http://utfit.roma1.infn.it/.
22. N. Isgur and M.B. Wise, Phys. Lett. B **232**, 113 (1989).
23. A. Giri, Y. Grossman, A. Soffer and J. Zupan, Phys. Rev. D **68**, 054018 (2003); A. Bondar, 2002 (unpublished).
24. A. Abulencia *et al.* [CDF Collab.], Phys. Rev. Lett. **97**, 242003 (2006).
25. W.S. Hou, Phys. Rev. D **48**, 2342 (1993).
26. K. Ikado *et al.* [Belle Collab.], Phys. Rev. Lett. **97**, 251802 (2006).
27. O. Tajima *et al.* [Belle Collab.], Phys. Rev. Lett. **98**, 132001 (2007).
28. M. Staric *et al.* [Belle Collab.], Phys. Rev. Lett. **98**, 211803 (2007).
29. B. Aubert *et al.* [BaBar Collab.], Phys. Rev. Lett. **98**, 211802 (2007).
30. A.F. Falk *et al.*, Phys. Rev. D **69**, 114021 (2004).

CP Violation: Experimental Results in B Decays

Livio Lanceri

Dipartimento di Fisica, Università di Trieste and INFN, Sezione di Trieste
livio.lanceri@ts.infn.it

1 Introduction

In a School devoted to "High Energy Physics: the Standard Model and Beyond" the physics of flavour and the violation of discrete symmetries in fundamental interactions are relevant in more than one way. While testing the Standard Model (SM) predictions, experiments in this sector can both provide a window to the discovery of new physics, and improve the knowledge of key ingredients of our present understanding of the evolution of the universe.

The present set of lectures gives a brief introduction to the phenomenology of *CP* violation in the context of the SM, concentrating then on a review of recent experimental results on *B* meson decays. This is a mature and well-documented field of research; both textbooks [1] and frequently updated in-depth reviews [2], [3] are now available: therefore, in this write-up the emphasis will be on highlighting the present main trends and directions, quoting a few relevant examples, with no attempt at a complete coverage of a very rich phenomenology.

These lectures share a substantial overlap with those previously given at another School [4]; experimental results are however updated at the time of writing them up.

2 Discrete Symmetries: Historical Perspective

Charge conjugation C changes each particle into its anti-particle; parity P reflects all space coordinates; time reversal T changes the sign of the time coordinate and, in quantum mechanics, also implies complex conjugation.

Nature seems to be symmetrical for combined *CPT* transformations. Indeed, no experimental test gave evidence of *CPT* violation up to now [5]. *CPT* invariance is guaranteed by a theorem for quantum field theories that satisfy Lorentz and space-time translation invariance, possess a lowest "vacuum" state in their energy spectrum, and obey "micro-causality" (commutation or anti-commutation relations between all distinct fields). As a direct consequence of this invariance, every particle has the same mass and lifetime as its anti-particle; their magnetic moments are equal in size, and have opposite directions. Due to *CPT* invariance, *CP* violation implies *T* violation and vice-versa.

Electromagnetic and strong interactions are both *C*- and *P*-symmetric. On the contrary, weak interactions are known to maximally violate both the *C* and *P* symmetry, while retaining an approximate symmetry for the combined *CP* transformation, as apparent for instance in an analysis of neutrino or anti-neutrino handedness in charged pions decays.

CP violation was first observed in 1964 in the neutral *K*-meson system [6]. The role of *CP* violation in cosmology was recognized as early as in 1967 by Sakharov [7], who included it among the necessary conditions for a dynamically-generated matter-antimatter asymmetry of our Universe.

In 1973 M. Kobayashi and T. Maskawa realized that *CP* violation can be accomodated in the renormalizable theory of weak interactions only if there are at least six quark flavours [8]. The first experimental evidence of the existence of quarks belonging to the third family was obtained in 1977, with the observation of a 9.5 GeV dimuon resonance in 400 GeV proton-nucleus collisions [9] and evidence for a close-by second resonance [10], later followed by a third [11]. Experiments at the DORIS-II electron-positron collider confirmed these narrow resonances and established their interpretation as radial excitations of the $b\bar{b}$ bound system [12].

A higher luminosity electron-positron collider, the Cornell Electron Storage Ring (CESR), was needed to observe the $\Upsilon(4S)$ fourth radial excitation at center-of-mass energy of about 10.58 GeV: its larger natural width was interpreted as indication that the $\Upsilon(4S)$ is above threshold for "open beauty" production [13]. Soon later, in 1983, a handful of hadronic decays of charged and neutral *b*-flavoured *B* mesons were exclusively reconstructed [14].

A few years later, in 1987, the mixing of neutral *B* mesons was established, with contributions from experiments at both proton-antiproton [15] and electron-positron [16] colliders.

The proposal to look for large *CP* violation effects in relatively rare *B* decays, based on the Kobayashi-Maskawa model, was put forward in 1980-81 by Carter, Sanda [17], and Bigi [18].

The following generation of experiments at the Tevatron proton-antiproton collider and at electron-positron colliders (LEP, SLC at the Z^0, CESR at the $\Upsilon(4S)$) made great progress in the understanding of heavy quark dynamics, collecting millions of events with *b*-hadrons and studying their properties, including some initial searches for *CP* violation. However, *CP* violation in *B* mesons was still beyond their statistical reach.

For this reason, in the mid 80's and early 90's several "*B* Factory" concepts were proposed with a production of about 10^7 *B*-meson pairs per year as a reference target. The projected performance of upgraded and future hadron colliders in the production and detection of the relevant B decay modes was also frequently discussed and compared with the electron-positron approach.

Finally two asymmetric-energy [19] high-luminosity electron-positron colliders, designed to operate at the $\Upsilon(4S)$ center-of-mass energy, were approved and built at SLAC (PEP-II [20]) and KEK (KEK-B [21]). About two years after the beginning of data taking at the two Factories, in 2002 the *BABAR* and BELLE Collaborations announced the discovery of "mixing-induced" *CP* violation in neutral B mesons [22]. In 2004 "direct" *CP* violation in the decays of charged B mesons was seen by both experiments [23, 24].

These discoveries and the developement of experimental methods to measure time-dependent *CP* asymmetries in B decays opened the way to a systematic investigation of the full pattern of *CP* violation effects predicted by the Standard Model in the quark sector. The following Sections are devoted to a discussion of these studies and of the opportunities they offer to find evidence for physics beyond the SM.

3 Phenomenology of *CP* Violation

The following Subsections introduce the basics of the phenomenology of *CP* violation. The connection with mixing in pseudoscalar neutral mesons is explored first. An outline of the CKM mechanism of the Standard Model follows. The main observables are finally reviewed.

3.1 Pseudoscalar Mesons

Since the first observation in the K system, neutral pseudoscalar mesons have been a good laboratory for the study of *CP* violation. Their time evolution resembles that of a classical system of two coupled oscillators with damping.

When one of the two oscillators is set in motion, energy is transferred to the other by their coupling; a back-and-forth energy transfer is established and is visible in the fact that the amplitudes of the individual motions themselves alternatively oscillate between a maximum and a minimum; the coupling strength determines the frequency of this superimposed back-and-forth overall oscillation. This peculiar motion can be described as a superposition of the two "normal modes" of the system, in which the individual oscillators steadily move exactly in phase or with opposite phase, at two typical slightly different frequencies. Damping affects these two modes differently: one of them dies away more quickly than the other. Waiting long enough, the two oscillators tend to move in phase with a vanishing amplitude.

The individual oscillators in this analogy correspond to the opposite "flavour eigenstates" of the particle-antiparticle system. Their opposite quantum numbers are conserved by strong interactions; they are not conserved by weak interactions, that are responsible both for the "coupling" (mixing) and the "damping" (decay) of the system. Normal modes correspond instead to the full weak Hamiltonian "mass eigenstates" or stationary states. The same formalism can be used for K^0, D^0, B^0 and B_s, collectively represented as P^0: only the main results are sketched here.

Oscillating and decaying neutral mesons can be described by a two-component wave function $\psi(t)$, combination of the flavour eigenstates $(P^0, \bar{P}^0)$.

In the Wigner-Weisskopf approximation [25], a two-component Schrödinger-like equation governs the evolution of $\psi(t)$ in proper time t:

$$i\frac{d\psi}{dt} = \mathbf{H}\psi = (\mathbf{M} - \frac{i}{2}\mathbf{\Gamma})\psi; \tag{1}$$

$\mathbf{H}$ is the weak effective hamiltonian in the $(P^0, \bar{P}^0)$ subspace, where the 2×2 matrices $\mathbf{M}$ and $\mathbf{\Gamma}$ are Hermitian, and are associated with transitions via off-shell (dispersive) and on-shell (absorptive) intermediate states, respectively. This is made explicit by expressing the matrices $\mathbf{M}$ and $\mathbf{\Gamma}$ in second-order weak-interactions perturbation theory, by sums over intermediate states. Diagonal elements of $\mathbf{M}$ and $\mathbf{\Gamma}$ are associated with the flavour-conserving transitions $P^0 \to P^0$ and $\overline{P}^0 \to \overline{P}^0$, while off-diagonal elements are associated with flavour-changing transitions $P^0 \leftrightarrow \overline{P}^0$.

Mass Eigenvalues and Eigenstates

One obtains the expressions for the mass eigenstates, in the original flavour basis, diagonalizing $\mathbf{H}$ and finding its complex eigenvalues:

$$\mu_{H,L} = m_{H,L} - (i/2)\Gamma_{H,L} \tag{2}$$

where the mass (H, heavy, and L, light) is here chosen to label the eigenstates, as is customary for the B system. The mass and lifetime differences of the two eigenstates are defined as:

$$\Delta m = m_H - m_L, \quad \Delta\Gamma = \Gamma_H - \Gamma_L. \tag{3}$$

Dimensionless parameters x and y are also usually defined as:

$$x \equiv \frac{\Delta m}{\Gamma}, \; y \equiv \frac{\Delta\Gamma}{2\Gamma}; \quad m \equiv \frac{m_H + m_L}{2}, \; \Gamma \equiv \frac{\Gamma_H + \Gamma_L}{2} \tag{4}$$

The mass difference Δm, or equivalently the x parameter, reflects the coupling that drives the oscillation frequency between flavour eigenstates, while $\Delta\Gamma$ (or y) shows the different effect of "damping" on the mass eigenstates, leading to different lifetimes for them.

Assuming *CPT* invariance in the following, mass eigenstates can then be written, in the flavour basis:

$$|P_H\rangle = p|P^0\rangle + q|\bar{P}^0\rangle, \quad |P_L\rangle = p|P^0\rangle - q|\bar{P}^0\rangle. \tag{5}$$

The magnitudes of the coefficients p and q are close to $1/\sqrt{2}$, so that the mass eigenstates do not differ much from the *CP* eigenstates $|P_{\pm}\rangle \equiv (1/\sqrt{2})(|P^0\rangle \pm |\bar{P}^0\rangle)$, and would coincide with them in case of *CP* conservation. A small violation can be expressed through the *CP*-violating parameter $\tilde{\varepsilon} \equiv (p-q)/(p+q)$. The mixing parameters p and q are determined by the off-diagonal matrix elements of $\mathbf{M}$ and $\mathbf{\Gamma}$. This can be shown introducing the difference of eigenvalues:

$$\Delta\mu = \Delta m - \frac{i}{2}\Delta\Gamma = 2\sqrt{\mathbf{H}_{12}\mathbf{H}_{21}}, \tag{6}$$

and deriving the following relation:

$$\frac{q}{p} = \frac{\Delta\mu}{2M_{12} - i\Gamma_{12}} = \sqrt{\frac{2M_{12}^* - i\Gamma_{12}^*}{2M_{12} - i\Gamma_{12}}}. \tag{7}$$

By definition $|p|^2 + |q|^2 = 1$; if CP is violated, a non-zero parameter δ can also be defined:

$$\delta = |p|^2 - |q|^2 = \langle P_L|P_H\rangle, \tag{8}$$

so that:

$$|p|^2 = \frac{1+\delta}{2}, \quad |q|^2 = \frac{1-\delta}{2}. \tag{9}$$

In terms of the off-diagonal matrix elements of $\mathbf{M}$ and $\mathbf{\Gamma}$, the *CP*-violating parameter δ is found to be:

$$\delta = \frac{2\mathrm{Im}(M_{12}^*\Gamma_{12})}{\Delta m^2 + |\Gamma_{12}|^2}. \tag{10}$$

Time Evolution, Amplitudes and Rates

The mass eigenstates $P^0_{H(L)}$ have exponential evolution as a function of proper time t, with well-defined masses and decay widths. As a consequence, an initially pure $P^0(\overline{P}^0)$ flavour eigenstate, that contains both mass eigenstates, will evolve to a superposition including both flavour components, in proportions that depend on proper time t. One can then derive the time-dependent decay rates for mesons initially tagged as flavour eigenstates, to final states such as semileptonic decays, that are self-flavour tagging. In the approximation $\Gamma_H = \Gamma_L = \Gamma$ for B mesons, these mixing probabilities with initial flavour-tag are simply expressed as:

$$|g_{\pm}|^2 = \frac{e^{-\Gamma t}}{2}[1 \pm \cos(\Delta m t)]. \tag{11}$$

The decay rates of mesons initially flavour-tagged as P^0 ($\overline{P}^0$) to a specific final state $f \neq \bar{f}$ can be obtained from the time-evolved flavour-eigenstates, squaring the resulting time-dependent amplitudes. An interesting special case is the decay rate of initially flavour-tagged B mesons to a common *CP* eigenstate $f = f_{CP}$, in the approximation of equal decay widths for the two mass eigenstates $\Gamma_H = \Gamma_L = \Gamma$:

$$\frac{\mathrm{d}\Gamma[B^0_{\text{phys}}(t) \to f]}{\mathrm{d}t} \tag{12}$$
$$= \frac{|A_f|^2 e^{-\Gamma t}}{2} \left\{[1 + \cos(\Delta m t)] + [1 - \cos(\Delta m t)]|\lambda_f|^2 - 2\mathrm{Im}\lambda_f \sin(\Delta m t)\right\}$$
$$\frac{\mathrm{d}\Gamma[\bar{B}^0{}_{\text{phys}}(t) \to f]}{\mathrm{d}t}$$
$$= \frac{|A_f|^2 e^{-\Gamma t}}{2} \left\{[1 - \cos(\Delta m t)] + [1 + \cos(\Delta m t)]|\lambda_f|^2 + 2\mathrm{Im}\lambda_f \sin(\Delta m t)\right\},$$

where in each expression we can identify, in this order, the contributions of "direct decays", "mixing followed by decay", and "mixing-decay interference". The *CP* parameter λ_f combines the ratio of the two flavour-eigenstate decay-amplitudes with the ratio of the mixing coefficients:

$$\lambda_f = \frac{q}{p}\frac{\bar{A}_f}{A_f}. \tag{13}$$

From a measurement of time-dependent asymmetries between flavour-tagged rates, both the absolute value and the phase of the λ_f parameter can be determined. A more detailed discussion of the signatures of *CP* violation in the relevant observables will be given in Subsect. 3.3.

Mixing: Pseudoscalar Meson Peculiarities

The same formalism can be used to describe the time evolution of all neutral pseudoscalar mesons. However, their behaviour is experimentally quite different, in ways that can be accounted for by the Standard Model.

Table 1 summarizes the known decay and mixing parameters [3, 38], of $K^0 = (\bar{s}d)$, $D^0 = (c\bar{u})$ $B^0 = (\bar{b}d)$, and $B_s = (\bar{b}s)$, where the same units (ps^{-1}) are used for Γ and Δm, setting $\hbar = c = 1$.

Table 1. Lifetime and mixing parameters for neutral pseudoscalar mesons $K^0 = (\bar{s}d)$, $D^0 = (c\bar{u})$ $B^0 = (\bar{b}d)$, and $B_s = (\bar{b}s)$. Natural units (ps^{-1}) are used for Γ and Δm, setting $\hbar = c = 1$.

	$K^0/\overline{K^0}$	$D^0/\overline{D}^0$	$B^0/\overline{B}^0$	$B_s/\overline{B}_s$
τ (ps)	89.3 ± 0.1; 51700 ± 400	0.4101 ± 0.0015	1.530 ± 0.009	$1.437^{+0.031}_{-0.030}$
Γ (ps^{-1})	5.61×10^{-3}	2.438 ± 0.009	0.641 ± 0.016	0.696 ± 0.015
$y = \Delta\Gamma/2\Gamma$	-0.99664	$0.0066^{+0.0021}_{-0.0020}$	0.009 ± 0.037	$0.121^{+0.083}_{-0.090}$
Δm (ps^{-1})	$(5.301 \pm 0.014) \times 10^{-3}$		0.507 ± 0.005	$17.77 \pm 0.10 \pm 0.07$
$x = \Delta m/\Gamma$	0.945 ± 0.002	$0.0087^{+0.0030}_{0.0034}$	0.776 ± 0.008	

The neutral K system has the unique feature of very different lifetimes for the two mass eigenstates, that are therefore labelled S (short-lived) and L (long-lived); from $\Gamma_S \approx 579\Gamma_L$ it follows that $\Delta\Gamma \approx -\Gamma_S$ and $\Gamma \approx \Gamma_S/2$; moreover $\Gamma_S \approx 2\Delta m$, fourteen orders of magnitude smaller than the average mass; in terms of the dimensionless parameters, $x \approx -y \approx 1$. A second related characteristic feature is the dominance of the two-pion decay channel $CP = +1$, that has a much larger kinematically allowed phase space than any other decay channel and is therefore associated with the short-lived K^0_S, that in the limit of *CP* conservation coincides with the $CP = +1$ eigenstate ($\varepsilon_K \approx 0$, Subsect. 3.1).

The neutral B system is characterized by very similar lifetimes for the two mass eigenstates ($\Gamma_L \approx \Gamma_H$: $y \approx 0$) and by an oscillation frequency of the same order of the average lifetime ($x \approx 0.7$). A much higher oscillation frequency was expected and recently measured ($x \approx 25$) for B_s mesons [26]; the experimental value for the lifetime difference, that has still a rather large uncertainty, is compatible with the expectation $y \approx -0.1$. Finally, very small values were expected for the mixing parameters of the D^0 system, and experimental evidence for mixing was obtained only very recently [27].

The present knowledge of *CP* violating parameters cannot be summarized as concisely as the mixing behaviour. The corresponding observables are discussed and classified in Sect. 3.3.

CP violation was first seen as a small percentage of two-pion ($CP = +1$) decays of long-lived K^0_L, presently characterized by $|\varepsilon_K| = (2.28\pm0.02)\times10^{-3}$, compatible with being originated in mixing ($\Delta S = 2$ transitions). It is also visible in self-tagging semileptonic decays, showing a small asymmetry in the K^0 ($\overline{K}^0$) content of K^0_L ($\delta_L = (3.27\pm0.12)\times10^{-3}$). A challenging experimental program, pursued during three decades, established the presence of a small amount of direct violation ($\Delta S = 1$ transitions), quantified by the parameter $\text{Re}(\varepsilon'/\varepsilon) \approx \varepsilon'/\varepsilon = (1.72 \pm 0.18) \times 10^{-3}$.

CP violation has been observed also in B decays, not yet in B_s or D meson decays. Recent results for B mesons are discussed in Sect. 4; their interpretation in the context of the Standard Model is recalled in Sect. 5. Results for the K system are reviewed elsewhere [28] and will not be further discussed here: only a few comments on future perspectives are given in Sect. 6.

3.2 *CP* Violation in the Standard Model

From the connection between *CP* and T violation and the anti-unitary character of the T operator, *CP* violation can be tracked down to the presence of complex phase differences among amplitudes contributing coherently to a particular transition. One can classify such phases as "weak" or *CP*-odd (with opposite sign for the *CP*-conjugate process), and "strong" or *CP*-even (with the same sign for the *CP*-conjugate process).

Weak phases originate from complex couplings in the Lagrangian. In the SM with no mass terms for neutrinos, this is the result of Yukawa interactions and the structure of charged-current weak interactions in the quark sector, resulting in the Feynman rules for the quark vertices with a charged W boson, that include the Cabibbo-Kobayashi-Maskawa (CKM) matrix element $V_{\alpha k}$ connecting an up-type quark ($\alpha = u, c, t$) with a down-type quark ($k = d, s, b$).

The CKM Matrix

With three generations of fundamental fermions, the 3×3 CKM matrix $V_{\alpha k}$, including the unitarity constraint, can be parameterized with four parameters: three Euler angles and one phase. Alternatively, one can introduce parameterizations satisfying unitarity only approximately, but incorporating some experimental information. After the experimental observation in 1983 that the bottom quark decays predominantly to the charm quark: $|V_{cb}| \gg |V_{ub}|$, Wolfenstein [29] noticed that $|V_{cb}| \sim |V_{us}|^2$ and proposed to use the Cabibbo angle $|V_{us}| = \lambda \sim 0.22$ as an expansion parameter:

$$\begin{pmatrix} V_{ud} & V_{us} & V_{ub} \\ V_{cd} & V_{cs} & V_{cb} \\ V_{td} & V_{ts} & V_{tb} \end{pmatrix} \simeq \begin{pmatrix} 1-\lambda^2/2 & \lambda & A\lambda^3(\rho - i\eta) \\ -\lambda & 1-\lambda^2/2 & A\lambda^2 \\ A\lambda^3(1-\rho-i\eta) & -A\lambda^2 & 1 \end{pmatrix} + \mathcal{O}(\lambda^4). \quad (14)$$

The parameter A is of order unity, because experimentally $|V_{cb}| \sim |V_{us}|^2$; moreover, $|V_{ub}|/|V_{cb}| \sim \lambda/2$ and therefore ρ and η should be smaller than one. In this parameterization, the *CP*-violating phase is included in V_{td} and V_{ub}, unitarity is satisfied up to order λ^3, and the hierarchy of couplings is manifest in the orders of magnitudes of the matrix elements, that can be expressed as powers of λ.

For precision tests of the SM, "improved" Wolfenstein-like parameterizations have been adopted, including higher order corrections [30].

Unitarity Triangles

CP violation in the SM corresponds to a non-zero value for the Wolfenstein parameter η. The most effective way of comparing experimental results to SM expectations in the CKM sector is by means of the unitarity relations between CKM matrix elements, that can be represented by Unitarity Triangles (UT) in the complex plane. The amount of *CP* violation is linked to the equal areas $2J$ of these different triangles, where J is the 'Jarlskog invariant' [31]. When studying $B_u^+ = (\bar{b}u)$ and $B_d^0 = (\bar{b}d)$ meson decays, the relevant triangular relation is between the first and third column of the matrix:

$$V_{ub}^* V_{ud} + V_{cb}^* V_{cd} + V_{tb}^* V_{td} = 0. \quad (15)$$

When normalized to $V_{cb}^* V_{cd}$, this relation corresponds to the rescaled triangle shown in Fig. 1; its Unitarity Angles or relative weak phases are:

$$\alpha \equiv \phi_2 \equiv \arg\left[-\frac{V_{td}V_{tb}^*}{V_{ud}V_{ub}^*}\right] \quad \beta \equiv \phi_1 \equiv \arg\left[-\frac{V_{cd}V_{cb}^*}{V_{td}V_{tb}^*}\right] \quad \gamma \equiv \phi_3 \equiv \arg\left[-\frac{V_{ud}V_{ub}^*}{V_{cd}V_{cb}^*}\right] \tag{16}$$

The three sides of this UT have comparable sizes, all of the same order $\simeq \lambda^3$; they have been measured from decay and mixing rates. From experiment, assuming the CKM model, the three angles are all expected to be significantly different from zero and lead to the SM expectation of large *CP* asymmetries in some *B* decays.

In the improved Wolfenstein parameterization the apex of the rescaled UT (Fig. 1) is given by:

$$\overline{\rho} + i\overline{\eta} \equiv -\frac{V_{ud}V_{ub}^*}{V_{cd}V_{cb}^*}, \tag{17}$$

where $\overline{\rho} = \rho(1 - \lambda^2/2 + \mathcal{O}(\lambda^4))$ and $\overline{\eta} = \eta(1 - \lambda^2/2 + \mathcal{O}(\lambda^4))$. The expected constraints from different observables, assuming a common set of SM parameters and setting to zero the uncertainties, are represented as lines in the $(\overline{\rho}, \overline{\eta})$ plane in Fig. 1 b [32]. In these lectures we are mostly concerned with measurements of the unitarity angles at the *B* Factories: sin2β, sin2α, $\sin\gamma$. Other measurements with *B* mesons constrain the sides of the triangle: $|V_{ub}|/|V_{cb}|$ is obtained from semileptonic decays without and with charm; $|V_{td}|$ from $B\overline{B}$-mixing. Also shown are the lines corresponding to ε_K and possible future observables ($K^+ \to \pi^+\nu\overline{\nu}$, $K^0 \to \pi^0\nu\overline{\nu}$) from the *K*-meson sector.

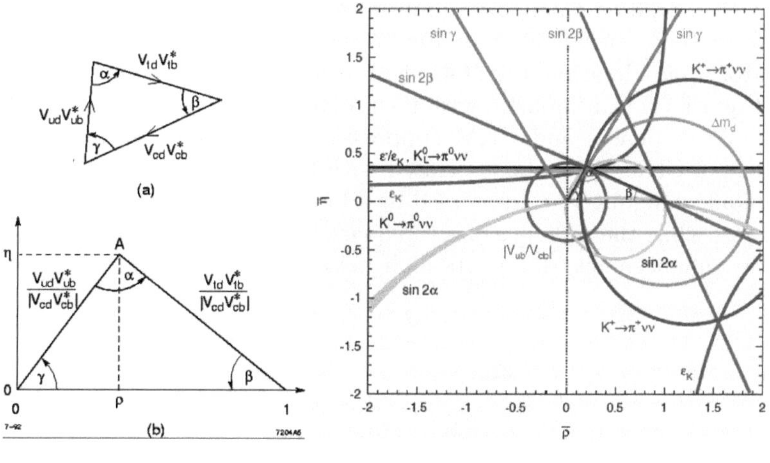

Fig. 1. (top) (a) Representation in the complex plane of the Unitarity Triangle formed by the CKM matrix elements $V_{ud}V_{ub}^*$, $V_{td}V_{tb}^*$, and $V_{cd}V_{cb}^*$. (b) Rescaled triangle, all sides divided by $V_{cd}V_{cb}^*$, with vertices A, B, and C at $(\overline{\rho}, \overline{\eta})$, $(1, 0)$, and $(0, 0)$, respectively. (bottom) Experimental paths to the Unitarity Triangle: a compilation of the theoretically expected constraints on the $(\overline{\rho}, \overline{\eta})$ apex; in this paper we are mostly concerned with measurements of $\sin 2\beta$, $\sin 2\alpha$, $\sin\gamma$. A common set of SM parameters is assumed here for all expected constraints, represented with their ambiguities (for instance two straight lines from the $\sin 2\beta$ measurement), but without theoretical or experimental uncertainties; the figure is reproduced from [32].

Mixing and Decay Amplitudes

The analysis of *CP*-violating observables for the extraction of the corresponding SM parameters is based on the difference of weak (that is, *CP*-odd) phases carried by the CKM matrix elements in different amplitudes that contribute to the same transition; some observables also depend on the relative magnitudes of the amplitudes or on differences of strong phases.

In the SM mixing is described at the constituent level by box diagrams. For B_d mesons, the matrix element M_{12} is dominated by "box diagrams" with a top quark in the loop (Fig. 2); Γ_{12} receives contributions from only a few common on-shell states, with total branching fraction of the order of 10^{-3}. As a result, $|\Gamma_{12}| \ll |M_{12}|$, and the following approximate expressions hold:

$$\Delta m \approx 2|M_{12}|, \quad \Delta\Gamma \approx \frac{2\mathrm{Re}(M_{12}\Gamma_{12}^*)}{|M_{12}|} \ll \Delta m, \quad \frac{q}{p} \approx -\frac{|M_{12}|}{M_{12}}. \tag{18}$$

In particular for $B_d = (\bar{b}d)$ the difference $\Delta\Gamma$ between decay widths of the heavy and light mass eigenstates is expected to be very small, in contrast to the large difference observed in the neutral K system. To a good approximation $|q/p| \simeq 1$, and the phase of q/p coincides with the phase of M_{12}. Finally, the CP-violating parameter δ is also expected to be small ($\sim 10^{-3}$) in the SM.

Hadronic B meson decays to two mesons can be described by the dominant tree-level and gluonic-penguin amplitudes corresponding to quark diagrams similar to those shown later in Fig. 7 for a specific decay channel. In the spectator approximation, the B meson is assumed to decay through the decay of its b quark (or $\bar{b}$ antiquark), with no intervention from the other "spectator" quark. Weak phases and CKM suppression factors are determined by the products of CKM elements associated with the quark-W vertices.

Gluonic penguins are suppressed by a loop factor with respect to tree-level diagrams, while the presence of a top quark in the loop partially compensates with an enhancement, due to its high mass. Penguin diagrams in which the

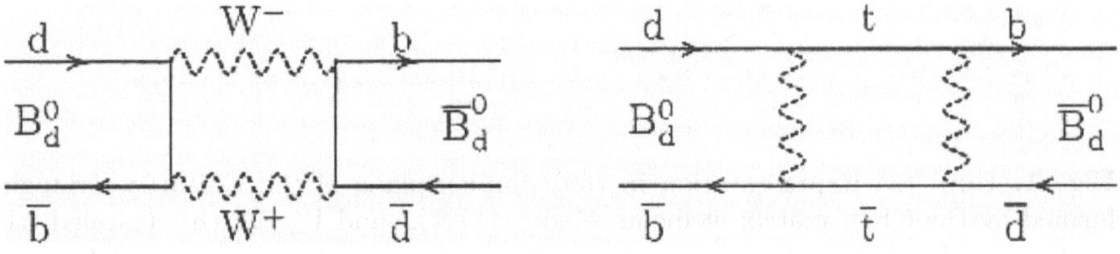

Fig. 2. The B_d^0 mass eigenstates can be expressed as $B_{L,H}^0 = pB^0 \pm q\bar{B}^0$. From the box diagrams describing the B^0-$\bar{B}^0$ transition amplitude M_{12}, it can be shown that, from V_{td} appearing twice in each diagram, $(q/p)_B \approx \sqrt{M_{12}^*}/\sqrt{M_{12}} = (V_{td}V_{tb}^*)/(V_{td}^*V_{tb}) = \exp(-i2\beta)$.

gluon is replaced by a Z boson or a photon are called electroweak (EW) penguins, and are suppressed by the smaller couplings.

B decays can also be mediated by diagrams involving the spectator quark: "W-exchange" and "penguin-annihilation" only contribute to neutral-meson decays, while the "W-annihilation" diagram only contributes to charged-meson decays. It is generally assumed that these diagrams are further suppressed, due to the smallness of the B wave function at the origin.

While some observables are directly linked to the CKM weak phases, in general the computation of amplitudes requires a correct treatment of the strong interactions among the bound quarks, and of final-state rescattering. A discussion of the corresponding theoretical techniques and of their limitations is outside the scope of this experimental review.

3.3 Observables

The violation of the *CP* symmetry can be observed in processes where amplitudes with different weak (*CP*-odd) phases contribute coherently, producing detectable interference effects.

CP Violation in the Decay

CP violation in the decay, or *direct CP* violation, can occur for both charged and neutral particles, when the total decay-amplitudes A_f and $\bar{A}_{\bar{f}}$ imply different rates for the process $P \to f$ and its *CP* conjugate $\bar{P} \to \bar{f}$ ($|A_f|^2/|\bar{A}_{\bar{f}}|^2 \neq 1$). The contribution of at least two interfering amplitudes with different weak *and* strong phases is required, namely $A_f = A_1 + A_2$ (Fig. 3) to produce an observable asymmetry

$$\begin{aligned} a_{CP} \equiv \frac{\Gamma(P \to f) - \Gamma(\bar{P} \to \bar{f})}{\Gamma(P \to f) + \Gamma(\bar{P} \to \bar{f})} &= \frac{|\bar{A}_{\bar{f}}/A_f|^2 - 1}{|\bar{A}_{\bar{f}}/A_f|^2 + 1} \\ &= \frac{2|A_1 A_2| \sin\phi \sin\delta}{|A_1|^2 + |A_2|^2 + 2|A_1 A_2| \cos\phi \cos\delta}, \end{aligned} \tag{19}$$

where ϕ (δ) is the difference in weak (strong) phase between the contributing amplitudes A_1 and A_2. According to the SM, in some rare B-meson decay channels ($\mathcal{B} \sim \mathcal{O}(10^{-5})$), mediated by suppressed tree-diagrams with sizeable penguin-diagram contributions asymmetries of order $\mathcal{O}(10\%)$ can be expected [33].

CP Violation in the Mixing

This type of violation is restricted to neutral mesons: it corresponds to mass eigenstates not being identical to *CP* eigenstates or, in terms of the mixing parameters, to $|q/p| \neq 1$. *CP* violation in the mixing of B mesons can be

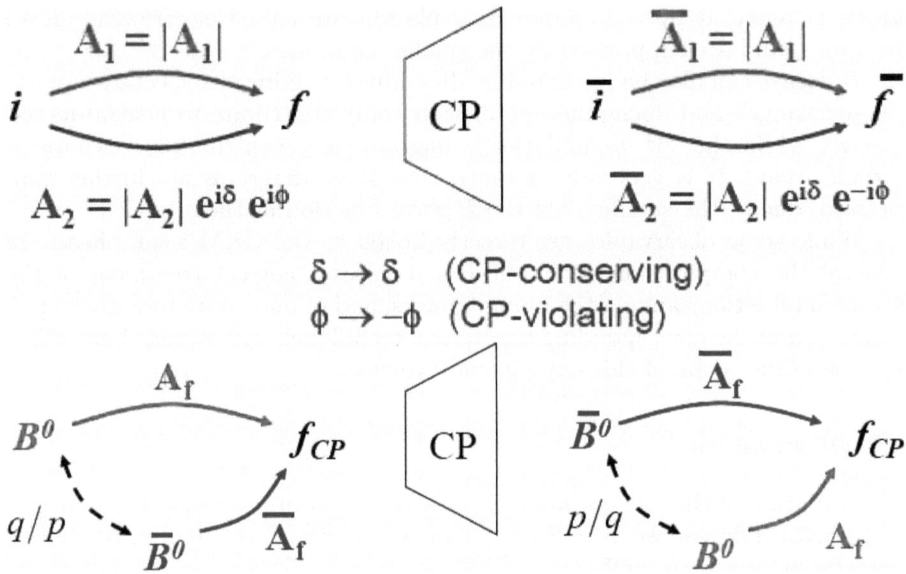

Fig. 3. Effect of the "*CP* mirror" on interfering decay amplitudes. (top) The direct *CP* asymmetry in the transition between an initial state i and a final state f is due to the interference between two amplitudes A_1 and A_2 with a relative *CP*-conserving phase δ and a *CP*-violating phase ϕ; (bottom) *CP* asymmetry in B^0 decay to a *CP* eigenstate f_{CP} from the interference between mixing, described by parameters p and q, and the decay amplitudes A_f and $\bar{A}_f$.

probed by comparing the probabilities of $B^0 \to \overline{B}^0$ and $\overline{B}^0 \to B^0$ and defining an asymmetry:

$$a \equiv \frac{P(\overline{B}^0 \to B^0) - P(B^0 \to \overline{B}^0)}{P(\overline{B}^0 \to B^0) - P(B^0 \to \overline{B}^0)} \tag{20}$$

In experiments at the $\Upsilon(4S)$, where $B^0\overline{B}^0$ pairs are produced, different rates of events with positive and negative same-sign dileptons:

$$B^0 \to \overline{B}^0 \to \ell^- X,\ \ \overline{B}^0 \to \ell^- X; \qquad B^0 \to \ell^+ X,\ \ \overline{B}^0 \to B^0 \to \ell^+ X, \tag{21}$$

would result in an asymmetry:

$$a_{SL} \equiv \frac{\Gamma(\Upsilon(4S) \to \ell^-\ell^-) - \Gamma(\Upsilon(4S) \to \ell^+\ell^+)}{\Gamma(\Upsilon(4S) \to \ell^-\ell^-) + \Gamma(\Upsilon(4S) \to \ell^+\ell^+)} = \frac{|p/q|^2 - |q/p|^2}{|p/q|^2 + |q/p|^2} = \frac{1 - |q/p|^4}{1 + |q/p|^4}. \tag{22}$$

In the SM, due to a very small deviation from one of $|q/p|$:

$$\left|\frac{q}{p}\right| \approx 1 - \mathrm{Im}\frac{\Gamma_{12}}{M_{12}}, \tag{23}$$

an asymmetry $a_{SL} \approx 10^{-3}$ is predicted [34].

Mixing-induced *CP* Violation

CP violation in neutral pseudoscalar mesons can also arise from the interference of mixing and decay amplitudes: this is also called "mixing-induced" violation. As shown in Fig. 3 (bottom), this type of violation leads to an observable time-dependent asymmetry $a_{f_{CP}}(t)$ in decays of flavour-tagged neutral B mesons to *CP* eigenstates f_{CP}:

$$a_{f_{CP}}(t) \equiv \frac{\Gamma[B^0_{\text{phys}}(t) \to f] - \Gamma[\bar{B}^0{}_{\text{phys}}(t) \to f]}{\Gamma[B^0_{\text{phys}}(t) \to f] + \Gamma[\bar{B}^0{}_{\text{phys}}(t) \to f]} \tag{24}$$

$$= \mathcal{C}_f \cos(\Delta mt) - \mathcal{S}_f \sin(\Delta mt), \tag{25}$$

$$\mathcal{C}_f = \frac{1 - |\lambda_f|^2}{1 + |\lambda_f|^2} = -\mathcal{A}_f, \quad \mathcal{S}_f = \frac{2\text{Im}\lambda_f}{1 + |\lambda_f|^2},$$

where for the cosine coefficient the notation $\mathcal{A}_f = -\mathcal{C}_f$ is preferred by the BELLE Collaboration. The asymmetry can be different from zero even in case of no direct violation ($|\bar{A}_f/A_f| = 1$) and no violation in the mixing ($|q/p| = 1$): in fact, if $|\lambda_f| = 1$, $a_{f_{CP}}$ becomes simply:

$$a_{f_{CP}}(t) = \text{Im}(\lambda_f) \sin(\Delta mt), \tag{26}$$

and measures the weak phase difference between the mixing and decay amplitudes.

4 *B* Mesons: Experimental Status

The search and discovery of *CP* violation in B mesons is an outstanding example of a crucial experimental test, succesfully performed after decades of ingenuous developement of new experimental facilities and techniques, on well-defined predictions made in the early days of the establishment of a theory: in this case the Kobayashi-Maskawa mechanism, incorporated in the Standard Model.

4.1 Experimental Methods

Experimentation on B decays has been successful both at hadron and electron-positron colliders. The results on *CP* violation were obtained at high-luminosity, asymmetric-energy electron-positron colliders operating just above the threshold for the production of $B\overline{B}$ pairs. Their properties and the main experimental ingredients of measurements of *CP* asymmetries are summarized in the following Subsections.

Colliders

The cross section for $b\bar{b}$ production in electron-positron colliders is much lower than the corresponding cross-section at hadron colliders. However, resonant production of B-meson pairs with high-luminosity electron-positron colliders running at the $\Upsilon(4S)$ center-of-mass energy is particularly effective, because the resonant $b\bar{b}$ cross-section of 1.2 nb, compared with the continuum $q\bar{q}$ background of 3.5 nb, gives a very good signal-to-background ratio of about 1 : 3. It is also very convenient because the $\Upsilon(4S)$ is just massive enough, with 10.580 GeV, to decay to a pair of $B_u^+ B_u^-$ or $B_d^0 \bar{B}_d^0$ mesons with a combined mass of 10.577 GeV, with no additional particles. This very clean environment is compatible with an inclusive trigger, retaining virtually all B-meson decays, and allows their exclusive reconstruction with high efficiency. There is a disadvantage as well, since the B's are produced with a momentum of about 345 MeV, and travel only about 30 μm on average, before decaying: this short distance cannot be experimentally resolved by the available detectors.

For *CP*-violation studies, that require the separation of B decay vertices for the measurement of time-dependent asymmetries, this drawback was offset by the clever idea of an asymmetric-energy [19] high-luminosity electron-positron collider, designed to operate at the $\Upsilon(4S)$ center-of-mass energy, with a Lorentz boost in the laboratory frame to amplify the separation of decay vertices and to make it detectable.

At present, two such facilities are operational, the PEP-II and KEKB B-Factories, together with the upgraded Tevatron proton-antiproton collider.

PEP-II at SLAC [20] collides 3.1 GeV positrons on 9.0 GeV electrons; it exceeded the design peak luminosity by about a factor three, reaching $1.2 \times 10^{34}\,\mathrm{cm}^{-2}\mathrm{s}^{-1}$. The total luminosity recorded by the *BABAR* detector was close to $400\,\mathrm{fb}^{-1}$ at the end of Summer 2006.

KEK-B at KEK [21] collides 3.5 GeV positrons on 8.0 GeV electrons; their peak luminosity, exceeding the design, reached the world-record value of $1.7 \times 10^{34}\,\mathrm{cm}^{-2}\mathrm{s}^{-1}$. A total integrated luminosity of more than $540\,\mathrm{fb}^{-1}$ was recorded by Summer 2006.

Both experiments devote about 10% of their running time to collecting off-resonance data, to measure the continuum background. The overall B-meson data sample, including both experiments, exceeds by now one billion pairs.

Detectors

The detector design for both *BABAR* and BELLE was optimized for the measurement of time-dependent *CP*-asymmetries. The three main ingredients are illustrated in Fig. 4: exclusive reconstruction of a specific B decay on one side, ("B_{rec}"), inclusive flavour tagging of the other B ("B_{tag}"), and measurement of their decay time difference Δt.

The *BABAR* and BELLE detectors, shown in Fig. 5, are described in detail elsewhere [35]; they share several common features, while differing in some of

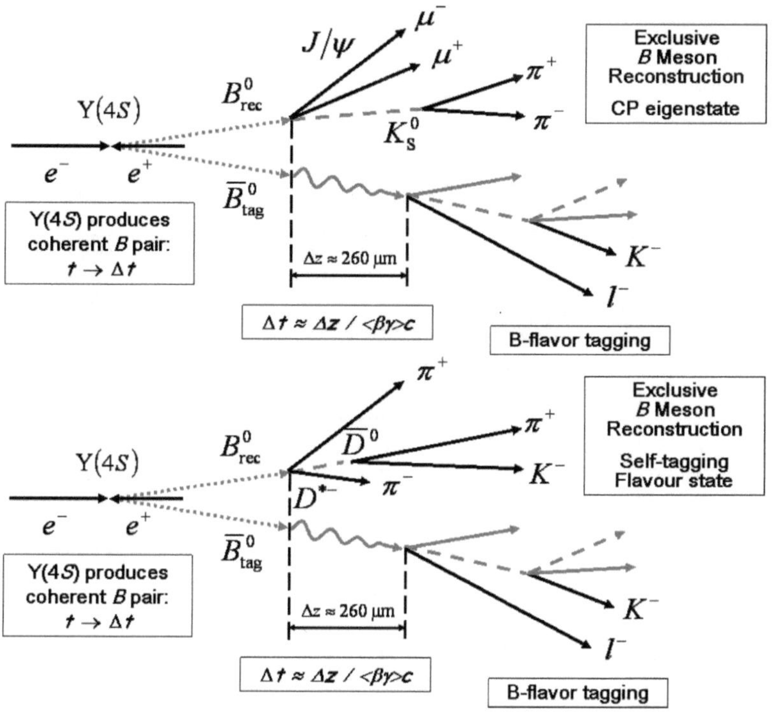

Fig. 4. The measurement of a *CP* asymmetry at a *B* Factory includes three ingredients (top figure): exclusive reconstruction of the decay to a specific (*CP* eigenstate) final state ("B_{rec}"), inclusive flavour tagging of the other B ("B_{tag}"), and measurement of their decay time difference Δt. The flavour-tagging efficiency and Δt resolution-function parameters are measured on a large control sample with exclusively reconstructed self-tagging final states ("B_{flav}") such as $D^{*-}\pi^+$ (bottom figure).

the adopted techniques. Both cover large fractions of the solid angle in the CM frame around the interaction region. Their multi-layer double-sided inner silicon microstrip detectors sample the charged particle tracks close to the interaction region and to the decay points, with space resolutions down to about 10–20 μm per point. The five-layer configuration adopted by *BABAR* is particularly effective also in the reconstruction of tracks from charged particles of low transverse momentum. Both experiments adopt a Drift Chamber as main tracking device to measure angle, momentum and energy deposition per unit path length of charged particle tracks. Charged hadron identification is obtained in *BABAR* by measuring the particle velocity by a novel Cherenkov ring-imaging device (DIRC, a Detector of Internally Reflected Cherenkov light), while BELLE adopted a more conventional combination of Time Of Flight (TOF) counters and aerogel threshold Cherenkov counters. Photon detection and electron identification are provided in both experiments

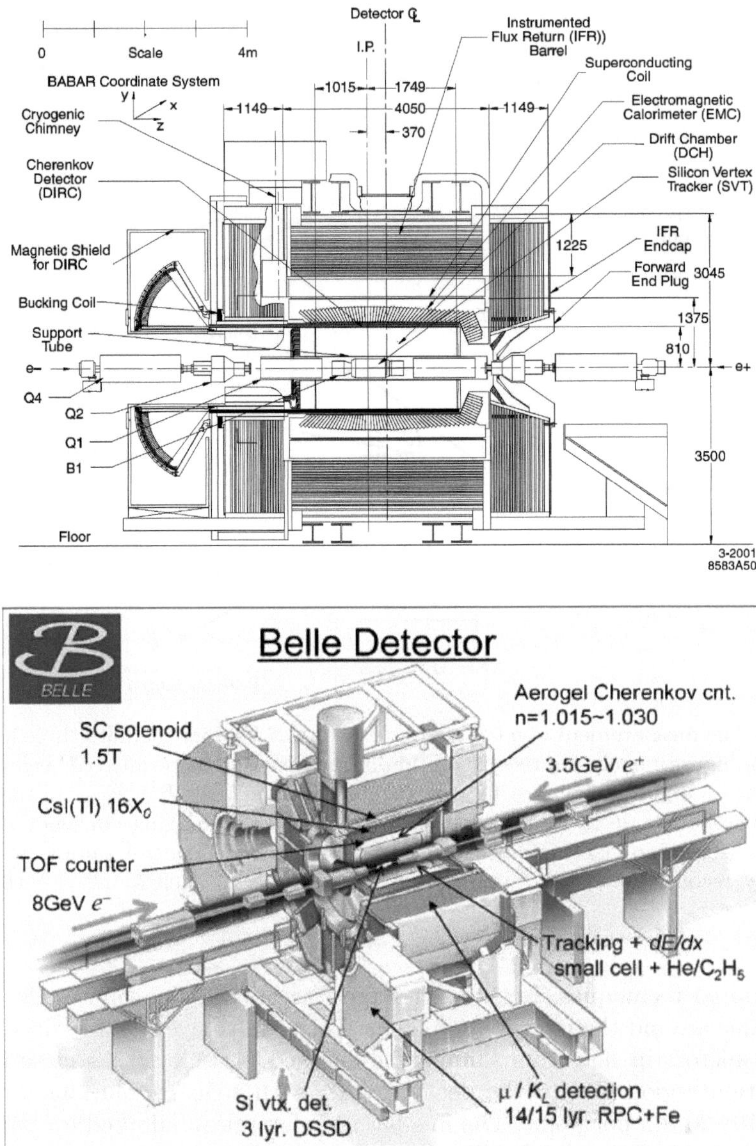

Fig. 5. The *BABAR* (top) and BELLE (bottom) detectors.

by electromagnetic calorimeters made of CsI(Tl) crystals. A superconducting solenoid provides the 1.5 T magnetic field for momentum analysis. Muons and neutral hadrons are identified by an external detector, using the iron that traps the return of the magnetic flux as a hadron filter, instrumented with Resistive Plate Chambers, partially substituted in *BABAR* by Streamer Tubes.

Analysis Techniques

Let us have now a closer look at the analysis methods for the determination of time-dependent *CP* asymmetries: exclusive reconstruction of specific B decays one one side, inclusive flavour tagging of the other B and measurement of their decay time difference Δt, and fits of the interesting physical parameters.

Exclusive reconstruction of B-candidates

One of the B mesons of the $B\overline{B}$ pair from the $\Upsilon(4S)$ resonance, produced almost at rest in the center-of-mass (CM) frame, is usually reconstructed exclusively, with efficiency typically in the 15−40% range, depending on the final state. The B candidates are reconstructed from tracks and energy clusters in the event; charged hadron identification is provided by the Cherenkov detectors and by the energy loss in the tracking detectors. Photon candidates are defined by energy clusters in the electromagnetic calorimeter, not associated with a charged track. Electron candidates are mainly identified by the ratio E/p of the energy deposit in the calorimeter to the momentum of the corresponding track. Muon candidates are primarily selected by their penetration depth in the instrumented iron yoke.

For all decay modes, except those involving K_L^0 and neutrinos, after reconstructing the decay chains including intermediate charm or other short-lived particles, two almost uncorrelated kinematical variables are computed to separate the signal from backgrounds. The variable ΔE is defined as the difference between the reconstructed energy of the B candidate in the e^+e^- CM frame and $\sqrt{s}/2$; the "beam-energy substituted-mass" is defined as $m_{ES} = \sqrt{(s/2 + \boldsymbol{p}_i \cdot \boldsymbol{p}_B)^2/E_i^2 - p_B^2}$; $\sqrt{s}$ is the total CM energy; the B momentum p_B and the four-momentum $(E_i, \boldsymbol{p}_i)$ of the e^+e^- initial state are defined in the laboratory frame. The m_{ES} resolution of about 2.6 MeV is dominated by the spread in beam energy, while the ΔE resolution (10–40 MeV, depending on the final state) is driven by detector effects, and in particular by energy resolution. Two typical examples of distributions of these variables are shown in Fig. 6.

Backgrounds are contributed by both continuum events and B-mesons.

Shape variables are used to discriminate the almost isotropically distributed decay products of B decays, from the jet-like configuration of particles in the background events due to continuum production of lighter $q\overline{q}$ pairs. In some cases, at a substantial cost in efficiency, an additional high-momentum lepton is required as a B-tag. Residual continuum backgrounds are statistically subtracted by using data recorded at a slightly lower CM energy ("off-resonance"), where the $\Upsilon(4S)$ does not contribute.

"Peaking" backgrounds in the signal region of the discriminating variables m_{ES} and ΔE may be contributed by partially mis-reconstructed B-mesons. In some cases, such as $\pi\pi$, πK or KK two-body decays, several interesting decay channels may constitute a background for each other. The detailed

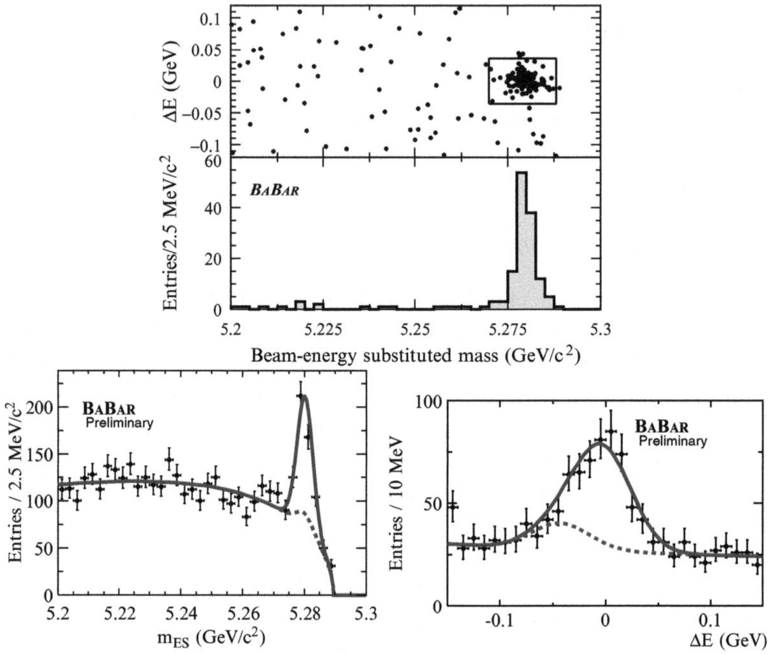

Fig. 6. Signal decays are identified using two kinematical variables: ΔE and m_{ES}, defined in the text. The first plot shows an example of the very clean $B^0 \to J/\psi K_S^0$ signal from the first year of data taking [36]; the other two plots are taken from a recently updated analysis of $B^0 \to \pi^+\pi^-$ [37], where the signal purity is lower.

knowledge of the detector particle-identification performance is used in such cases to separate the contributions with maximum-likelihood procedures.

Flavour tagging

For *CP* asymmetries, after the reconstruction of the relevant $B \to f$ final state, the second key ingredient is the tagging of the flavour of the other B. To maximize efficiency, this is achieved by an inclusive tag. Signatures include the correlated charge of fast leptons and kaons ($B^0 \to \ell^+ X$, $B^0 \to \overline{D}^0 \to K^+$), of soft pions from D^* decays, etc. Each tagging category has a discriminating power that can be described by the effective tagging efficiency $Q_i = \varepsilon_i(1-2w_i)^2$, where ε_i is the fraction of events associated with the tagging category i, and w_i is the probability of incorrectly assigning the tag. The statistical error on the measurements of *CP*-violating asymmetries and mixing are both proportional to $1/\sqrt{\sum_i Q_i} = 1/\sqrt{\varepsilon_{\text{eff}}}$.

The tagging efficiency ε_i and mistag fraction w_i for each category are measured on a large control sample of events with one B fully reconstructed in a self-tagging final state and the other providing the inclusive tag of type i. The amplitude of the observed time-dependent oscillation (equation (11)) is in fact diluted by the measurable factor $D_i = 1-2w_i$. *BABAR* relies on six tagging categories and uses a neural-network approach, with a combined effective efficiency $\varepsilon_{\text{eff}} = 0.305 \pm 0.004$. BELLE ranks events into categories based on mistag rates estimated by likelihood ratios, with a parameter ranging from zero (no discrimination) to one (unambiguous flavour assignment), obtaining a similar efficiency: $\varepsilon_{\text{eff}} = 0.287 \pm 0.005$.

Vertex reconstruction

The reconstruction of B-decay vertices is required to measure time-dependent *CP* asymmetries, where the time difference Δt is measured as $\Delta t \simeq \Delta z/(\gamma\beta c)$ from the space separation $\Delta z = z_{rec} - z_{tag}$ of the two B decay vertices along the boost direction. The Δz resolution, of the order of $170\,\mu$m, is dominated by the inclusively reconstructed B_{tag} side and is designed to be a fraction of the average separation $<\Delta z> \simeq \gamma\beta c\tau_B \simeq 250 \div 270\,\mu$m. The resolution functions are modelled using the B_{flav} control sample of fully reconstructed events, with inclusive tag. Typically, Δt resolutions are in the 1.1–1.4 ps range, where a lower value for *BABAR* is explained by the slightly higher boost at PEP-II.

CP asymmetry fits

Both *BABAR* and BELLE determine the relevant physical *CP* parameters using rather sophisticated fit procedures, that rely on data and control samples to characterize the detector efficiency, resolution and dilution factors.

As an example, in a likelihood fit aiming at the determination of the *CP* parameter λ_f, the signal portion of a sample of flavour-tagged candidate decays to a *CP* eigenstate f_{CP}, that we shall call f for brevity, is described (see Sect. 3.3) by the time-dependent probability distribution function (PDF) $\mathcal{P}_{\pm}^{CP}$:

$$\mathcal{P}_{\pm}^{CP}(\Delta t) = \frac{e^{-|\Delta t|/\tau_B}}{4\tau_B} \{1 \pm D\,[\mathcal{S}_f \sin \Delta m_d \Delta t - \mathcal{C}_f \cos \Delta m_d \Delta t]\} \otimes R(\Delta t), \tag{27}$$

that includes the experimental tagging dilution factors $D = 1 - 2w$ and the resolution function $R(\Delta t)$. The $\pm$ sign corresponds to the $B^0(\overline{B}^0)$ flavour tag, and the $\mathcal{S}_f$ and $\mathcal{C}_f$ coefficients are defined in terms of λ_f in eq.(24). Appropriate PDFs describe the background contributions to the selected sample of f candidates; their properties are studied on control samples and on simulated events.

The parameters describing efficiencies and dilution factors for different flavour-tagging categories, and the parameters of the resolution function R can be extracted in combined fits by using the significantly larger "flavour" sample; the corresponding signal PDF can be written as:

$$\mathcal{P}_{\pm}^{flav}(\Delta t) = \frac{e^{-|\Delta t|/\tau_B}}{4\tau_B} \{1 \pm D \cos \Delta m_d \Delta t\} \otimes R(\Delta t). \tag{28}$$

The "flavour" samples have also been used by both experiments for the determination of the B^0 lifetime τ_B and the mixing parameter Δm_d with improved sensitivity with respect to previous experiments.

The sensitivity in the measurement of the *CP* parameter $\mathrm{Im}(\lambda_f)$, assuming $|\lambda_f| = 1$, can be approximately obtained by a back-of-the-envelope calculation of the uncertainty on the *CP* asymmetry with a given number N_{obs} of observed signal events:

$$\delta a_f \simeq \frac{1}{D\sqrt{N_{\mathrm{obs}}}} = \frac{1}{D\sqrt{\varepsilon \times \mathcal{BF} \times N_{\mathrm{prod}}}}, \tag{29}$$

where the relevant figures of merit are the number of produced $B^0\overline{B}^0$ events $N_{\mathrm{prod}} = \int \mathcal{L}dt \times \sigma_{b\bar{b}} \times 2f_0$, that depends on the integrated luminosity, the $b\bar{b}$ cross-section, and the ratio of neutral to charged B-meson pair production f_0; the efficiency of data taking, *CP*-eigenstate reconstruction and tagging $\varepsilon = \varepsilon_{\mathrm{det}} \times \varepsilon_{\mathrm{CP}} \times \varepsilon_{\mathrm{tag}}$, the branching fraction $\mathcal{BF}$ of the reconstructed decay channel, and the dilution factors $D = d_{\mathrm{tag}} \times d_{\mathrm{bkg}}$, that include both mistagging $d_{\mathrm{tag}} = 1 - 2w$ and the background dilution effect $d_{\mathrm{bkg}} = N_S/(N_S + N_B)$; N_S and N_B are the signal and background rates respectively. The final uncertainty depends on careful optimization of the analysis and detailed background studies. Softer selection criteria increase efficiency at the cost of more accepted background; one obtains an overall sensitivity gain if the background components are characterized and described in the fit in terms of appropriate discriminating variables.

4.2 Limits on *CP* Violation in Mixing

The Standard Model predicts a very small *CP* violation in the mixing:

$$1 - \left|\frac{q}{p}\right|^2 \simeq \mathrm{Im}\left(\frac{\Gamma_{12}}{M_{12}}\right) \sim \mathcal{O}(10^{-3}). \tag{30}$$

The dilepton asymmetry a_{SL} defined in equation (22) can be measured to determine $|q/p|$, and the real part of the $\varepsilon_B = (p-q)/(p+q)$ parameter:

$$\begin{aligned} a_{SL} &\equiv \frac{\Gamma(\Upsilon(4S) \to \ell^-\ell^-) - \Gamma(\Upsilon(4S) \to \ell^+\ell^+)}{\Gamma(\Upsilon(4S) \to \ell^-\ell^-) + \Gamma(\Upsilon(4S) \to \ell^+\ell^+)} \\ &= \frac{1 - |q/p|^4}{1 + |q/p|^4} \simeq \frac{4\mathrm{Re}(\varepsilon_B)}{1 + |\varepsilon_B|^2}; \\ \left|\frac{q}{p}\right| &= \sqrt[4]{\frac{1 - a_{SL}}{1 + a_{SL}}} \end{aligned} \tag{31}$$

The experimental status, including older measurements by LEP experiments, and more recent results by CLEO, *BABAR* and BELLE at the $\Upsilon(4S)$, is summarized as follows in the Winter 2007 world-average by the HFAG group [38], combining statistical and systematic uncertainties:

$$\begin{aligned} a_{SL} &= -0.0049 \pm 0.0038 \\ |q/p| &= 1.0025 \pm 0.0019 \\ \frac{\mathrm{Re}(\varepsilon_B)}{1+|\varepsilon_B|^2} &= -0.0012 \pm 0.0010. \end{aligned} \tag{32}$$

The corresponding parameters in B_s mixing, from measurements at the Tevatron by CDF and D0, are also compatible with zero:

$$a_{SL} = -0.0030 \pm 0.0101 \tag{33}$$

$$|q/p| = 1.0015 \pm 0.0051 \tag{34}$$

Even in the clean B-Factory environment, measuring such small asymmetries is a challenge that requires a very good control of systematic effects.

4.3 Direct *CP* Violation

Direct *CP* violation, not involving particle-antiparticle oscillations, was observed in $K_L^0 \to \pi\pi$ decays [39], after a difficult search, that took several decades after the initial discovery of *CP* violation. In the K-meson system the effect is very small, due to the suppression of one of the interfering decay amplitudes, and to additional cancellations, that are to a large extent accidental. Conversely, if *CP* violation is due to the Kobayashi-Maskawa mechanism described in Sect. 3.2, large effects are possible and expected in some decay channels of the B-meson system.

A good candidate for the observation of the direct *CP* asymmetry (Sect. 3.3) is the decay $B^0 \to K^+\pi^-$, whose branching fraction is $(1.89\pm0.07)\times10^{-5}$ [38]: in the Standard Model, this decay occurs through two different amplitudes, tree and penguin. The tree amplitude ($b \to u\overline{u}s$), carrying the V_{ub} CKM factor, is strongly CKM-suppressed, and its weak phase is different with respect to the penguin amplitude ($b \to sq\overline{q}$). As a result all the conditions for large direct *CP* violation may be satisfied by the two interfering amplitudes: comparable magnitudes, different weak phases and, within large theoretical uncertainties, different strong phases.

Direct *CP* violation in the decay $B^0 \to K^+\pi^-$ was observed as a time-integrated asymmetry

$$A_{K\pi} \equiv \frac{n_{K^-\pi^+} - n_{K^+\pi^-}}{n_{K^-\pi^+} + n_{K^+\pi^-}} \tag{35}$$

by *BABAR* and confirmed by BELLE in 2004, with a combined significance exceeding five standard deviations. The present average result, including a measurement by CDF, is

$$A_{K\pi} = -0.095 \pm 0.013. \tag{36}$$

The experimental observation is indeed important because it supports the validity of the CKM mechanism without particle-antiparticle mixing, contributing to rule out *CP* violation models of the "superweak" type. It also helps disentangling penguins, trees, weak and strong phases.

No evidence of direct *CP* asymmetries is seen yet in other $K\pi$ modes, like $K^{\pm}\pi^0$ from charged $B^{\pm}$, where the experimental uncertainty is somewhat larger. Direct *CP* violation can also be searched for in the time-dependent B-decay asymmetries (Sect. 3.3); the experimental status of these searches will be discussed in Sect. 4.4.

4.4 Unitarity Triangle: Angles

Sizeable *CP* violating time-dependent asymmetries are expected by the CKM model in the interference of mixing and decay amplitudes of neutral B mesons to some final states f. As explained in Sect. 3.3, these asymmetries can be used to determine the weak phase difference between mixing and decay amplitudes, that are linked to the Unitarity Angles in the CKM model.

The Unitarity Angle $\beta(\phi_1)$

The Unitarity Angle β (ϕ_1), given by $-\arg V_{td}$ in the Wolfenstein parameterization, can be measured in several different ways in B-meson decays. The theoretically cleanest determination requires the measurement of time-dependent *CP* asymmetries, in decays of neutral B mesons to *CP* eigenstates (Sect. 3.3) with charmonium ($b \to c\bar{c}s$ modes). The measurements performed in "pure penguin" $b \to sq\bar{q}$ modes are appealing because they are sensitive to deviations due to New Physics, through virtual particles in the penguin-diagrams loop. Other processes, such as $b \to c\bar{c}d$, are considered separately, because they receive contributions from penguin diagrams with different weak phase.

$b \to c\bar{c}s$ modes

The decays of neutral B mesons to $J/\psi K^0_S$ and other channels mediated at quark level by $b \to c\bar{c}s$, are particularly clean: theoretically, because the dominating tree and the penguin decay amplitudes (Fig. 7) have the same weak phase relative to the mixing amplitude, so that the uncertainty of the prediction for the coefficient $\mathcal{S}_f = -\eta_f \sin 2\beta$ is very small, at the percent level; experimentally, because the final state has a relatively large branching fraction ($\mathcal{BF}(B \to J/\psi K^0) = (8.72 \pm 0.33) \times 10^{-4}$ [3]) and can be reconstructed with large efficiency and good purity.

For the reference channels $f = J/\psi K^0_{S,L}$ the *CP* parameter λ_f includes the effects of B mixing, B decay, and K mixing, resulting in an overall phase 2β:

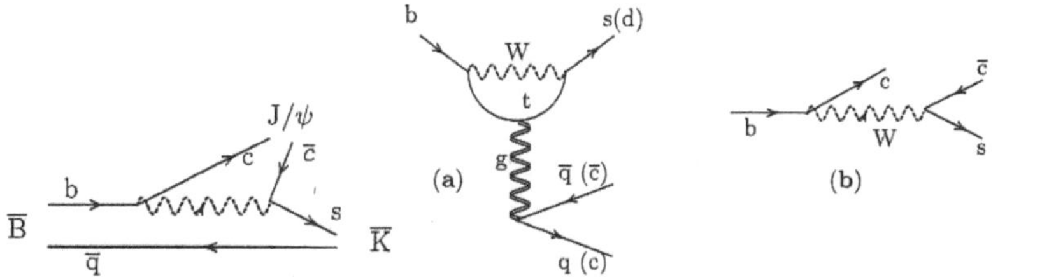

Fig. 7. (left) Colour-suppressed decay amplitude for the decay $\overline{B} \to J/\psi\overline{K}$. (right) The penguin (a) and tree (b) diagrams for the transition $b \to sc\bar{c}$ have the same weak phase.

$$\lambda_f = \left(\frac{q}{p}\right)_B \frac{\bar{A}_f}{A_f} = e^{-2i\beta}\eta_f \left(\frac{V_{cb}V_{cs}^*}{V_{cb}^*V_{cs}}\right)\left(\frac{p}{q}\right)_K = \qquad (37)$$

$$e^{-2i\beta}\eta_f \left(\frac{V_{cb}V_{cs}^*}{V_{cb}^*V_{cs}}\right)\left(\frac{V_{cs}V_{cd}^*}{V_{cs}^*V_{cd}}\right)$$

leading to a time-dependent asymmetry

$$a_f(\Delta t) = \mathrm{Im}\lambda_f \sin(\Delta m \Delta t) = -\eta_f \sin 2\beta \sin(\Delta m \Delta t) \qquad (38)$$

where the *CP*-parity η_f is $+1$ (-1) for the decays to K_L^0 (K_S^0).

In 2002 both *BABAR* and BELLE announced a positive result [22], establishing *CP* violation in the *B* system for the first time and confirming the prediction of the Standard Model. An example of the measured time-dependent asymmetry is shown in Fig. 8 together with a similar asymmetry measured in the *K* system; in both systems asymmetries can be quite large, involving however a small fraction of the total decay rate.

Since then, both experiments have kept updating their analyses, including more data and more decay modes. The Winter 2007 average value [38] from charmonium modes is

$$\sin 2\beta = 0.678 \pm 0.026, \qquad (39)$$

where the error reflects the combined statistical and systematic uncertainties; the combined statistical uncertainty alone is now ± 0.022 and is still the limiting factor. Figure 9 shows the corresponding constraints in the $(\overline{\rho}, \overline{\eta})$ plane.

The *BABAR* updated result [42] $\sin 2\beta = 0.714 \pm 0.032 \pm 0.018$ is based on a data sample of 383×10^6 $B\overline{B}$ events. It includes the decay modes $J/\psi K_S^0$, $\psi(2S)K_S^0$, $\chi_{c1}K_S^0$, $\eta_c K_S^0$ $(\eta_f = -1)$ and $J/\psi K_L^0$ $(\eta_f = +1)$, with a total of about 13000 selected events ("*CP* sample"), with purity ranging between 96% and 55%, depending on the decay mode. A much larger "flavour" control sample of about 124000 self-tagging events is selected to provide tagging efficiencies and resolution functions parameters. The similar updated result by BELLE [43] $\sin 2\beta = 0.642 \pm 0.031 \pm 0.017$, on a sample of 535×10^6 $B\overline{B}$ events, concentrates on the $J/\psi K_S^0$ and $J/\psi K_L^0$ decay modes, with signal yields of 7484 ± 87 and 6512 ± 123 events respectively. For both experiments the $\mathcal{C}_f = -\mathcal{A}_f$ coefficient of the cosine term in the time-dependent asymmetry

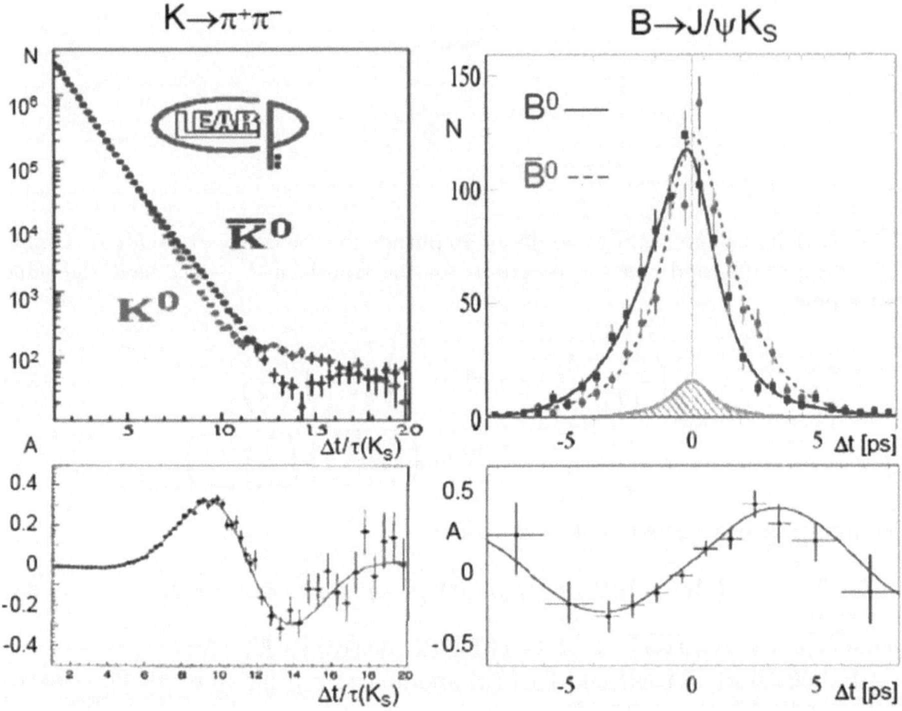

Fig. 8. Examples of observed *CP* asymmetries in decays to *CP* eigenstates of neutral K (CP-LEAR experiment [40], left) and neutral B-mesons (B-Factories: BABAR experiment [22], right). The top plots show the flavour-tagged time-dependent rates; The time difference $\Delta t = t_{CP} - t_{tag}$ between the decays of the two B mesons can be negative, depending on which of them occurs first; the shaded area (plot on the right) shows the small background contribution. The asymmetry shown in the bottom plots can be quite large in both cases, but it involves only small fractions of the total decay rates. For neutral K mesons the interference effect originating the asymmetry is maximal for decay times in the 9-15τ_S range, where τ_S is the K_S^0 lifetime. The raw asymmetry shown for $B \to J/\psi K_S^0$ decays, implies a maximum asymmetry of about 0.7 after corrections for experimental dilutions, but the corresponding branching fraction is smaller than 10^{-3}.

is compatible with zero (no direct *CP* violation) and statistical uncertainties in $\sin 2\beta$ are still larger than systematics.

These results are in good agreement with SM expectations based on the assumption of the validity of the CKM mechanism and on the previously available experimental constraints on the sides of the Unitarity Triangle, as shown in Sect. 5. In particular, the two bands representing the *CP* violation observations in K decays (ϵ_K) and in B decays ($\sin 2\beta$) do overlap with the UT apex region predicted by the intersection of bands from measurements of the UT sides ($|V_{ub}|$, $|V_{cb}|$ and B mixing). It should also be noted that

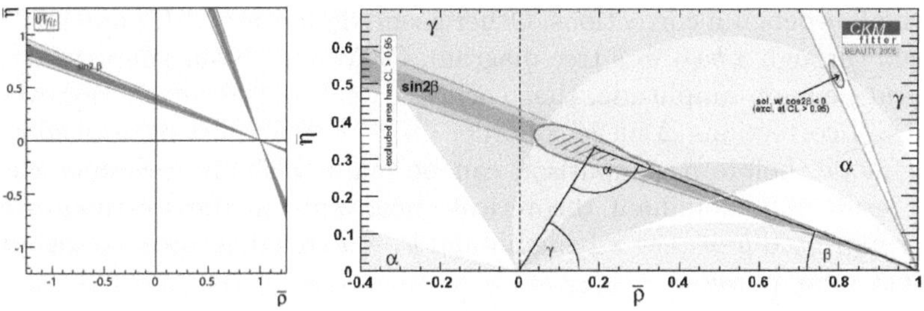

Fig. 9. Allowed region in the $(\overline{\rho}, \overline{\eta})$ plane derived from the experimental determinations of $\sin 2\beta$: as obtained (left) by the UTfit group [41] using $B^0 \to J/\psi K^0$ results only, and (right) by the CKMfitter group [32]. In the second case also the constraints from measurements of the other two unitarity angles are superimposed.

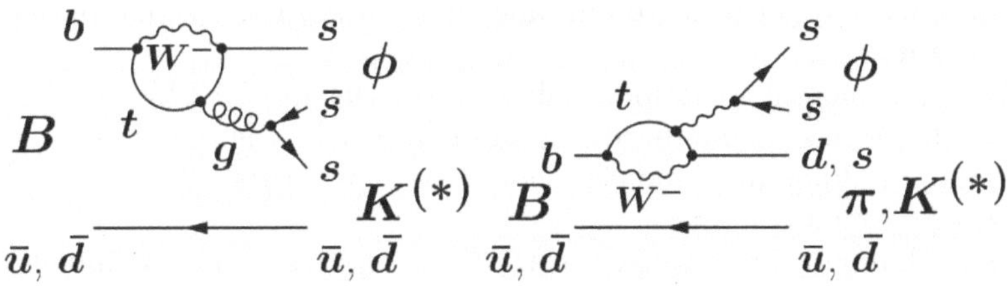

Fig. 10. Diagrams describing the decays $B \to \phi K$, ϕK^*, and $\phi\pi$: the dominant gluonic penguin (left), flavour singlet (right). In the latter, the $s\bar{s}$ pair may be connected to the loop either through gluons or an electroweak boson.

the measurement of $\sin 2\beta$ implies a four-fold ambiguity in the solutions for β. Both experiments developed methods to reduce the number of acceptable solutions, by complementary measurements: results are in agreement with the SM.

$b \to sq\overline{q}$ modes

After the confirmation of the validity of the CKM mechanism the interest is now focussed on $b \to s\bar{s}s$ decays, that receive contributions purely from "internal" and "flavour-singlet" penguin diagrams (Fig. 10), and are believed to be sensitive to New Physics through new virtual particle contributions in the loops.

The phase in the *CP* violation parameter $\lambda_{\phi K^0_S}$ from mixing-decay interference in $B \to \phi K^0_S$ (with $\eta_{CP} = -1$), is expected to be the same as in $\lambda_{J/\psi K^0_S}$, from $(q/p)_B = \exp(-2i\beta)$, so that the measurement of the asymmetry coefficient S_f should give the the same result for both $b \to c\bar{c}s$ and $b \to ss\bar{s}$, in the absence of New Physics effects.

The study of this channel, the cleanest both theoretically and experimentally, can be extended by looking at ϕK^0_L and $K^0_S K^0_S K^0_S$, both with $\eta_{CP} = -1$

and small u-penguin corrections. Other useful decays are $\pi^0 K_S^0$ and ωK_S^0, that proceed through a $\bar{b} \to \bar{u}u\bar{s}$ tree diagram, CKM and colour suppressed, and a $\bar{b} \to \bar{s}d\bar{d}$ penguin amplitude, that carries a weak phase $\sin 2\beta$. However, different SM corrections $\Delta \sin 2\beta$, of order $\mathcal{O}(\lambda^2) \sim 5\%$, must be applied to each decay mode, before a comparison can be made with the reference charmonium result. The dominant theoretical uncertainty in these estimates comes from suppressed $b \to u \to s$ penguin and $b \to u$ tree diagrams, contributing a different weak phase.

The measurement of time-dependent *CP* asymmetries to extract the $\mathcal{S}_f$ and $\mathcal{C}_f = -\mathcal{A}_f$ coefficients for these penguin modes is a challenging experimental task, that is reflected in larger statistical and systematic uncertainties. The branching fractions are in fact smaller by one or two orders of magnitude, and for some channels the decay chains include several neutral particles in the final state, with larger backgrounds and significantly lower signal purity in the selected samples.

The experimental situation, including *BABAR* and BELLE results shown at the 2007 Winter conferences, is summarized in Fig. 11. The $-\eta_f \times \mathcal{S}_f$ coefficient for different channels, should be equal to $\sin 2\beta$, up to the SM corrections discussed above.

For all channels the "effective" $\sin 2\beta^{\text{eff}} = -\eta_f \times \mathcal{S}_f$ is now within about one standard deviation from the $b \to c\bar{c}s$ value; it should be noted, however, that for all it is smaller than the reference charmonium value, while SM corrections

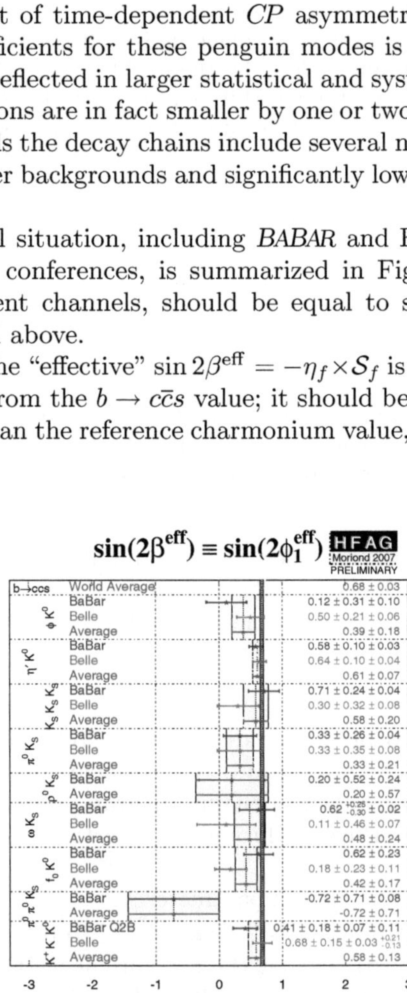

Fig. 11. Summary of $\sin 2\beta^{\text{eff}}$ measurements (HFAG, Winter 2007 averages) [38], from the $\mathcal{S}_f$ coefficient for charmonium (reference value of $\sin 2\beta$, narrow band in the plot) and several penguin-mediated channels, multiplied by the η_f *CP*-parity of the final state: $-\eta_f \times \mathcal{S}_f = \sin 2\beta^{\text{eff}}$.

favor larger values instead. This fact can still be taken as a hint at a possible non-SM effect, suggesting that more data are needed to narrow down the statistical uncertainties that dominate these measurements.

$b \to c\bar{c}d$ modes

Due to possible significant penguin-diagram contributions, the cosine and the sine coefficients of the Cabibbo-suppressed $b \to c\bar{c}d$ decays can be different from zero and $\sin 2\beta$ respectively, although the difference is not expected to be very large.

The time-dependent asymmetry measurements performed in these modes by both BELLE and *BABAR* are referenced in [38]. These results are largely dominated by statistical uncertainties: for a significant comparison with theoretical expectations larger data samples are needed.

The Unitarity Angle $\alpha(\phi_2)$

The measurement of the unitarity angle α (ϕ_2) is more challenging than the determination of β (ϕ_1).

The interference between mixing and decay to a final *CP* eigenstate involving a $b \to u$ transition, for example $B \to \pi^+\pi^-$ (Fig. 12, tree amplitude) provides sensitivity to $\alpha = \pi - (\beta + \gamma)$ through the *CP* parameter $\lambda_{\pi\pi}$:

$$\lambda_{\pi\pi} = \left(\frac{q}{p}\right)_B \frac{\bar{A}_{\pi\pi}}{A_{\pi\pi}} = e^{-2i\beta} \frac{\bar{A}_{\pi\pi}}{A_{\pi\pi}}. \tag{40}$$

In the absence of penguin contributions $\bar{A}_{\pi\pi}/A_{\pi\pi} = e^{-2i\gamma}$ and eq. (40) would be reduced to $\lambda_{\pi\pi} = e^{2i\alpha}$, with $\mathcal{S}_{\pi\pi} = \sin 2\alpha$ and $\mathcal{C}_{\pi\pi} = 0$. For $B \to \pi^+\pi^-$, however, a sizeable penguin contribution is expected, with $|\mathcal{P}_{\pi\pi}/\mathcal{T}_{\pi\pi}| \sim 0.3$;

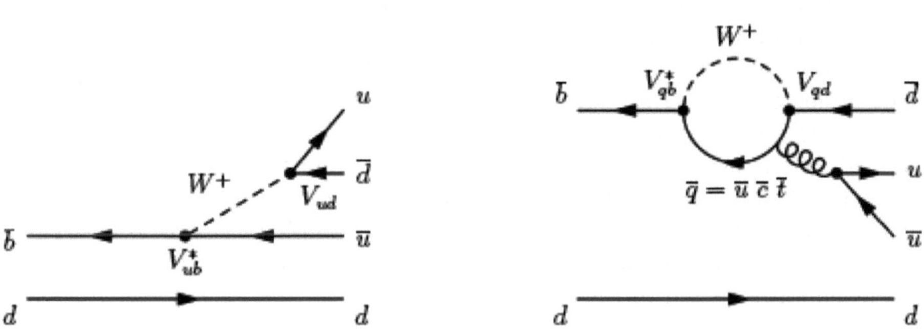

Fig. 12. Decay tree and penguin diagrams for $B^0 \to \pi^+\pi^-$: the corresponding amplitudes have the same order $\mathcal{O}(\lambda^3)$. The tree amplitude is proportional to $V_{ub}V_{ud}^*$ with weak phase $\gamma \simeq \arg V_{ub}$, while the phase of the penguin amplitude, proportional to $V_{tb}V_{td}^*$, is related to $\beta \simeq -\arg V_{td}$.

different weak phases are carried by the tree amplitude $\mathcal{T}_{\pi\pi}$ and by the penguin amplitude $\mathcal{P}_{\pi\pi}$, resulting in:

$$\lambda_{\pi\pi} = e^{i2\alpha} \frac{e^{-i\gamma} + \mathcal{P}_{\pi\pi}/\mathcal{T}_{\pi\pi}}{e^{+i\gamma} + \mathcal{P}_{\pi\pi}/\mathcal{T}_{\pi\pi}} \equiv |\lambda_{\pi\pi}| e^{2i\alpha_{\text{eff}}} . \tag{41}$$

As a consequence, the measurement of the time-dependent *CP* asymmetry defined in (24) determines the coefficients $\mathcal{S}_{\pi\pi}$ and $\mathcal{C}_{\pi\pi}$:

$$\mathcal{S}_{\pi\pi} = \sqrt{1 - \mathcal{C}^2_{\pi\pi}} \sin 2\alpha_{\text{eff}}, \quad \mathcal{C}_{\pi\pi} \propto \sin\delta, \tag{42}$$

where δ is the strong phase difference between the two contributing decay amplitudes. The effective α_{eff} value obtained from this measurement must be corrected for the deviation due to this so-called "penguin pollution", in order to obtain α.

The correction $\Delta\alpha = \alpha_{\text{eff}} - \alpha$ can be determined [44] by measurements of flavour-tagged decay rates of channels related by isospin transformations: for the $\pi\pi$ final state, these are $\pi^+\pi^-$, $\pi^+\pi^0$ and $\pi^0\pi^0$. The corresponding decay amplitudes enter triangular relations, from which $\Delta\alpha$ can be extracted.

Complementary approaches to extract α are based on measurements of time-dependent asymmetries of $B \to \rho\rho$, $\rho\pi$.

$B \to \pi\pi$

Extracting the unitarity angle $\alpha(\phi_2)$ from the *CP* asymmetry in the $\pi\pi$ decay channel is also an experimental challange, because of the small branching fractions of the relevant decay modes (world averages [38]: $\mathcal{BF}(B^0 \to \pi^+\pi^-) = (5.2 \pm 0.2) \times 10^{-6}$, $\mathcal{BF}(B^0 \to \pi^0\pi^0) = (1.31 \pm 0.21) \times 10^{-6}$, and $\mathcal{BF}(B^+ \to \pi^+\pi^0) = (5.7 \pm 0.4) \times 10^{-6}$). The main backgrounds for the $\pi^+\pi^-$ channel are continuum events with pairs of tracks from opposite jets, in the B-candidate invariant mass window, and $B \to K\pi$ decays, that have a larger branching fraction, with K misidentified as π. If these backgrounds are kept under control by a good particle identification system, the analysis is a "straightforward" application of the same methods developed for other time-dependent *CP* asymmetry measurements. The signal yields in recent *BABAR* and BELLE analyses reached 1139 ± 49 and 1464 ± 65 events in data sets of 383 and 535×10^6 $B\overline{B}$ pairs respectively. An example of the measured time-dependent asymmetry and the experimental values of the fitted $\mathcal{S}_{\pi\pi}$ and $\mathcal{C}_{\pi\pi}$ coefficients are shown in Fig. 13.

$B \to \rho\rho$

The smallness of the measured $\rho^0\rho^0$ branching fraction ($(1.07 \pm 0.38) \times 10^{-6}$ [46]) relative to the corresponding charged modes ($(23.1 \pm 3.3) \times 10^{-6}$ [38]) implies a smaller $\Delta\alpha_{\rho\rho}$ correction, to be obtained with the isospin method

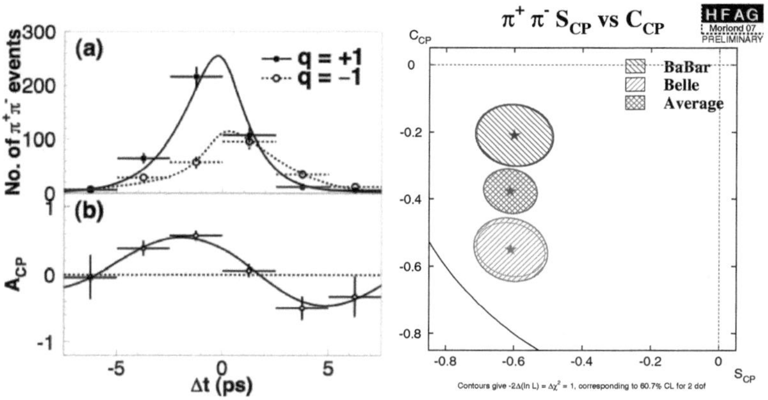

Fig. 13. (left) An example of time-dependent *CP* asymmetry measured in $B \to \pi^+\pi$ decays [45]. (right) Fitted values of the $\mathcal{S}_{\pi\pi}$ and $\mathcal{C}_{\pi\pi} = -\mathcal{A}_{\pi\pi}$ coefficients (S_{CP} and C_{CP} in the figure) from *BABAR* [37] and BELLE [45]. The discrepancy in the cosine coefficient should be resolved with more data.

outlined above. The vector-vector final state turns also out to be almost completely longitudinally polarized, so that the standard time-dependent *CP*-asymmetry analysis can be applied without a combined angular analysis, to obtain the averaged "longitudinal" coefficients [38]:

$$\mathcal{S}_{\rho\rho,\mathrm{long}} = -0.06 \pm 0.18 \quad \mathcal{C}_{\rho\rho,\mathrm{long}} = -0.11 \pm 0.13. \tag{43}$$

For the decay channel $B \to \rho^+\rho^- \to \pi^+\pi^0\pi^-\pi^0$ the four final particles have lower momenta than those in two-body final states, and include two neutral pions. Backgrounds are predominantly from continuum $q\overline{q}$, but include also a sizeable fraction of partially misreconstructed B decays, that overlap with the signal and may have different decay-time resolution properties: this makes the analysis rather challenging. Signal yields of 615 ± 57 and 576 ± 53 were obtained by *BABAR* and BELLE in data sets of $(347 \pm 3.9) \times 10^6$ and 535×10^6 $B\overline{B}$ pairs respectively.

$B \to \rho\pi$

The decay $B^0 \to (\rho\pi)^0 \to \pi^+\pi^-\pi^0$ offers the opportunity to directly extract both the weak and strong phases by a time-dependent analysis of the three-body final state [47], dominated by ρ resonances and their interference. Unlike $\pi^+\pi^-$, the final state $\rho^\pm\pi^\mp$ is not a *CP* eigenstate, and has a more complicated amplitude structure. For a model-independent extraction of the unitarity angle α a generalization of the isospin analysis described for the $\pi\pi$ channel implies pentagonal relations in this case [48]. The ultimate sensitivity to α in this mode is obtained by a time-dependent analysis extended to the entire region of interest in the Dalitz plot, that contains the ρ resonances and their interference [47]. The complex decay amplitudes and the *CP* violating weak phase α are simultaneously determined. The experiments determine 27 coefficients of the form factors bilinears from the fit to the data; physics parameters including the weak phase are determined from subsequent fits to these coefficients.

The analysis of $B \to (\rho\pi)^0 \to \pi^+\pi^-\pi^0$ decays is based on 2067 ± 86 signal events for *BABAR* [49], selected in a sample of $(375 \pm 4) \times 10^6$ $B\overline{B}$ pairs; and 971 ± 42 for BELLE [50] in 449×10^6 pairs. Also in this case the dominant background from $q\overline{q}$ continuum is discriminated by a combination of event shape variables, with algorithms tuned on off-resonance data; Monte Carlo simulations are used for detailed studies of other background sources from B decays, and of the fraction of misreconstructed signal decays.

Constraints on $\alpha(\phi_2)$

A detailed account of the asymmetry parameters extracted for each of these channels by the two experiments and their averages can be found elsewhere [38].

The combined constraints on α that can be derived from $b \to u\bar{u}d$ transitions are consistent with the CKM fits excluding these measurements from the inputs: combining the isospin analyses of $\pi\pi$, $\rho\rho$ and the Dalitz analysis of $\rho\pi$ the following result is obtained [38], using the fitting methods of ref. [32]:

$$\alpha = (100.2 \,^{+15}_{-8.8}\,)^\circ . \tag{44}$$

A similar analysis [41], with a different fitting method, including the only latest available data in $B \to \pi\pi$, $\rho\rho$, gives two solutions; one of them is compatible with the SM:

$$\alpha = (92 \pm 7)^\circ . \tag{45}$$

The α-range ruled out by $B \to \pi\pi$ is rather limited; most of the sensitivity is coming from the $B \to \rho\rho$ analysis, and $B \to \rho\pi$ is relevant in constraining the mirror solution allowed by $\rho\rho$. The corresponding combined constraints in the $(\overline{\rho}, \overline{\eta})$ plane are shown in Fig. 14.

The Unitarity Angle $\gamma(\phi_3)$

The direct determination of the unitarity angle γ poses the challenge of measuring the relative phase of a CKM-suppressed $b \to u$ transition with respect to the dominant $b \to c$ transition.

Figure 15 shows that two features allow $B^- \to D^0K^-$ and $B^- \to \overline{D}^0K^-$ decays to be used for this purpose: neutral D^0 and $\overline{D}^0$ mesons can decay to a common final state, and both neutral D's flavours are produced via $b \to c\bar{u}s$ and $b \to u\bar{c}s$ transitions. The relative phase betwen the interfering amplitudes is the difference $\delta - \gamma$ of strong and weak phases; for the charge conjugate B^+ decay the relative phase is the sum $\delta + \gamma$. The interference term,

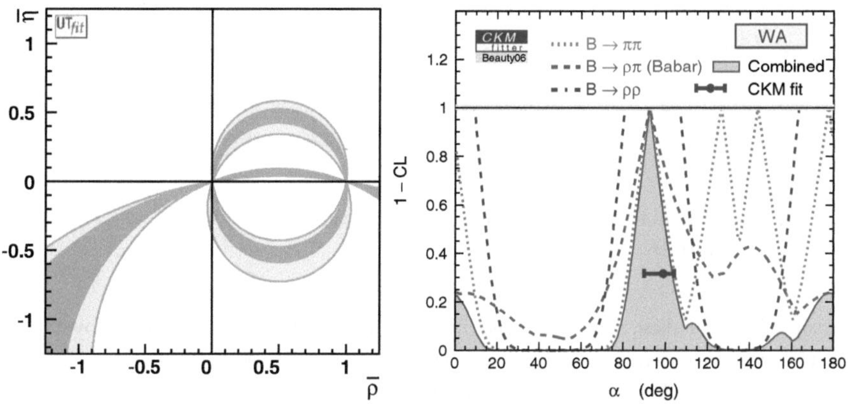

Fig. 14. (left) Combined constraints on $\alpha(\phi_2)$ from $B \to \pi\pi$, $B \to \rho\rho$, $B \to \rho\pi$ delimit an allowed region in the $(\bar{\rho}, \bar{\eta})$ plane [41]. (right) Confidence level as a function of α for each mode and for their combination [32]. The allowed value for α from direct measurements agrees with the result of the CKM global fit, obtained excluding these measurements from the fit inputs.

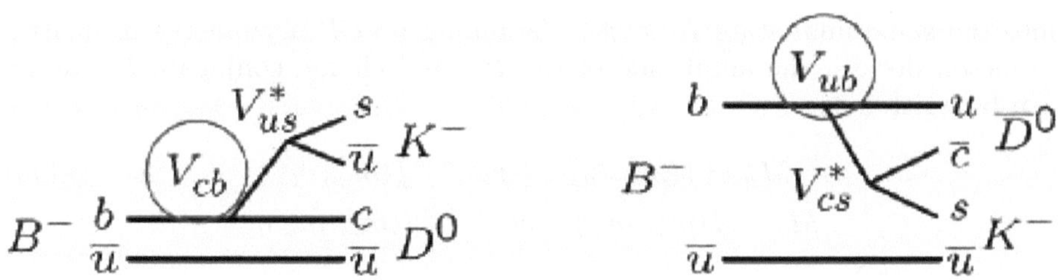

Fig. 15. Amplitudes for $B^- \to D^0 K^-$ and $B^- \to \overline{D}^0 K^-$.

sensitive to γ, that produces an observable asymmetry between B^- and B^+ decays, is suppressed by a factor $r_b \equiv A(b \to u)/A(b \to c) = R_u F_{cs}$ where $R_u = \sqrt{\bar{\rho}^2 + \bar{\eta}^2} \sim 0.4$ is the left side of the normalized Unitarity Triangle, and $F_{cs} \sim$ is a not well-known "colour suppression" factor [51], expected to be in the range $0.2 \div 0.5$.

Additional suppression factors in useful event rates, and therefore in the experimental sensitivity to γ, are due to the restrictions on the final states accessible from both D^0 and $\overline{D}^0$.

In the Gronau-London-Wyler (GLW) approach [52], these are the relatively rare decays to *CP* eigenstates such as K^+K^-, $\pi^+\pi^-$.

The Atwood-Dunietz-Soni (ADS) method [53] compensates for the r_b suppression factor by considering the "Cabibbo favored" $D^0 \to K^+\pi^-$, the "doubly Cabibbo suppressed" $\overline{D}^0 \to K^+\pi^-$ and the charge conjugate decays. In this case, the favoured $(b \to c)$ B decay followed by the doubly-suppressed D decay interferes with the suppressed $(b \to u)$ B decay followed by the favored D decay. The resulting combined decay amplitudes are smaller,

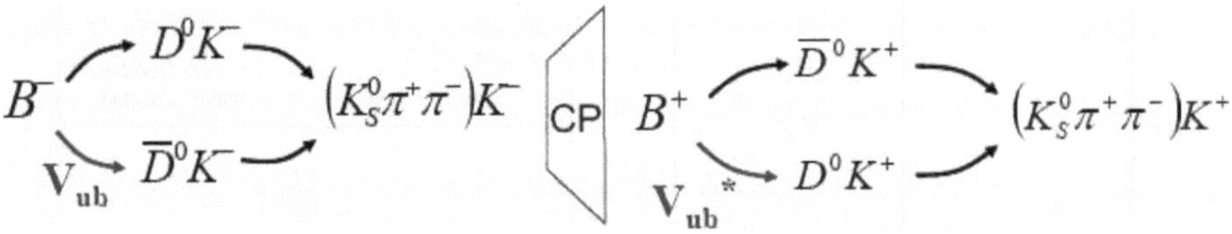

Fig. 16. Interference obtained using three-body decays $D^0, \overline{D}^0 \to K_S^0\pi^+\pi^-$ in $B^- \to D^0K^-$ and $B^- \to \overline{D}^0K^-$.

implying lower rates, but the possible *CP* asymmetry is enhanced by the fact that the combined amplitudes have the same order of magnitude.

For both the GLW and ADS methods, the present experimental uncertainties are still very large. At the cost of some model dependence, a better sensitivity is obtained [54] by using the three-body decays $D^0, \overline{D}^0 \to K_S^0\pi^+\pi^-$. The relative phase γ enters via an amplitude proportional to V_{ub}, as shown in Figs. 15 and 16.

In this analysis method [54] the amplitudes $M_1 \propto V_{cb}^* V_{us}$ (for $\overline{D}^0K^+$) and $M_2 \propto V_{ub}^* V_{cs}$ (for D^0K^+) interfere as the D^0 and $\overline{D}^0$ mesons decay into the same final state $K_S^0\pi^+\pi^-$. Assuming no *CP* asymmetry in neutral *D*-meson decays, the amplitude of the B^+ and charge conjugate B^- decay can be written as

$$
\begin{aligned}
M_+ &= f(m_+^2, m_-^2) + re^{i\gamma+i\delta} f(m_-^2, m_+^2) \\
M_- &= f(m_-^2, m_+^2) + re^{-i\gamma+i\delta} f(m_+^2, m_-^2),
\end{aligned}
\tag{46}
$$

where m_+^2 and m_-^2 are the squared invariant masses of the $K_S^0\pi^+$ and $K_S^0\pi^-$ combinations respectively, $f(m_+, m_-)$ is the complex amplitude for the $\overline{D}^0 \to K_S^0\pi^+\pi^-$ decay, and the absolute value of the ratio between the two interfering amplitudes, r, is predicted to be in the range 0.1-0.2. The functional form of f is fixed by a $\overline{D}^0 \to K_S^0\pi^+\pi^-$ decay model, with parametres obtained from a large sample of flavour-tagged D decays produced in continuum e^+e^- annihilation.

A simultaneous fit can then be performed for the B^+ and B^- decays with r, γ, and δ as free parameters. The parameter r is positive definite and needs a correction for an intrinsic positive bias if directly extracted from the fit; it is also positively correlated with the value of γ. A different set of less correlated variables is therefore used:

$$
\begin{aligned}
x_\pm &= \mathrm{Re}(re^{i(\delta\pm\gamma)}) = r\cos(\delta \pm \gamma) \\
y_\pm &= \mathrm{Im}(re^{i(\delta\pm\gamma)}) = r\sin(\delta \pm \gamma).
\end{aligned}
\tag{47}
$$

The *CP* violating asymmetry related to γ is apparent in the separation of (x_+, y_+) from (x_-, y_-).

A complete summary of the experimental results with the different methods for measuring γ is given in [38]: at present, the best sensitivity is obtained

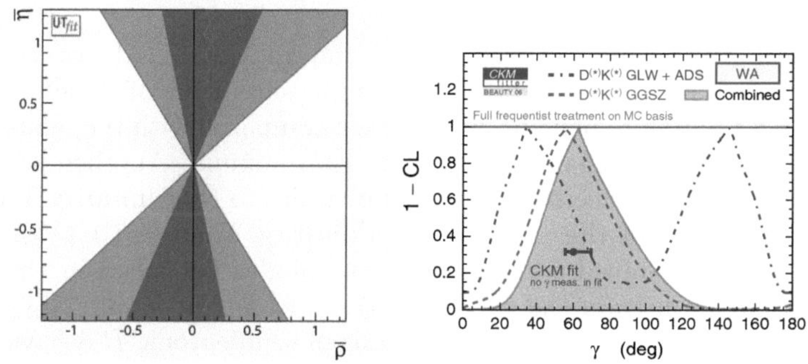

Fig. 17. (left) Combined constraints on $\gamma(\phi_3)$ from different measurements (Winter 2007 average) delimit an allowed region in the $(\bar{\rho}, \bar{\eta})$ [41]. (right) Combined confidence level as a function of γ (Summer 2006 [32]), in good agreement with the value obtained from a global fit, excluding the direct measurements of γ from the fit input.

with the "Dalitz" method. The direct determination from the combination of the three methods and using all the available data gives

$$\gamma = (83 \pm 19)^\circ \tag{48}$$

(Winter 2007, UTfit [41]) and

$$\gamma = (62^{+38}_{-24})^\circ \tag{49}$$

(Summer 2006, CKMfitter [32]). Figure 17 shows the overall allowed region in the $(\overline{\rho}, \overline{\eta})$ plane.

A different experimental trade-off is involved in another method [55] that uses the more frequent ($BF \simeq \mathcal{O}(10^{-3})$) decays $B^0 \to D^{\pm}\pi^{\mp}$, $B^0 \to D^{*\pm}\pi^{\mp}$ and $B^0 \to D^{\pm}\rho^{\mp}$, where small ($\mathcal{A}_{CP} \simeq \mathcal{O}(10^{-2})$) *CP* asymmetries are expected. The interference in this case is between the Cabibbo-favoured amplitude (e.g. $B^0 \to D^-\pi^+$) with the doubly Cabibbo-suppressed amplitude (e.g. $B^0 \to D^+\pi^-$). The relative weak phase between these two amplitudes is γ and, when combined with the B mixing phase, the total phase difference is $(2\beta + \gamma)$. The size of the time-dependent *CP* asymmetry is proportional to the ratio of magnitudes of the suppressed and favoured amplitudes, expected to be about 0.02, that can in principle be obtained experimentally from the corresponding suppressed charged B decays using isospin, or from self-tagging neutral B decays into strangeness using SU(3) symmetry. The feasibility of this method to measure γ was demonstrated [56, 57]; its sensitivity is limited experimentally by the smallness of the measured asymmetries and theoretically by the need of assumptions on the ratio of suppressed and favoured amplitudes, that cannot be measured directly.

5 Unitarity Triangle: Global Fits

One possible approach to tests of the Standard Model and searches for New Physics was discussed in Sect. 4.4, where measurements of the effective parameter $\sin 2\beta_{\rm eff}$ in pure penguin decays were compared with the reference value obtained from decay channels involving charmonium. Another possible way to look for non-SM effects is through global fits to the Unitarity Triangle.

Measurements of the sides of the Unitarity Triangle ($|V_{cb}^* V_{cd}|$, $|V_{ub}^* V_{ud}|$ and $|V_{tb}^* V_{td}|$), combined with those of the angles described in the previous Sections, over-constrain the Unitarity Triangle. In particular, $|V_{cb}|$ and $|V_{ub}|$ are determined by the branching fractions of semileptonic B decays with and without charmed hadrons in the final state, and $|V_{td}|$ is determined by B mixing. Complementary information is obtained from other B decays ($B \to \tau\nu$, $B \to \rho\gamma$) and from B_s mixing, measured by CDF [26] at the Tevatron.

Among the fits performed by several authors [58], we consider those of the CKMfitter [32] and UTfit [41] groups.

The constraints in the $(\bar{\rho}, \bar{\eta})$ plane, derived only from the measurements of the *CP*-violating parameters $\sin 2\beta(\phi_1)$, $\alpha(\phi_2)$ and $\gamma(\phi_3)$ in B-meson decays, are well matched to the constraint derived from the *CP* parameter ε_K measured in the neutral K-meson system. By now they are already giving a better determination of the Unitarity Triangle apex, than the "*CP* conserving" quantities related to the Unitarity Triangle sides: the measurements of $|V_{ub}/V_{cb}|$, Δm_d, Δm_s, and $\mathcal{BF}(B \to \tau\nu)$.

The global fits including the constraints from both angles and sides of the Unitarity Triangle are shown in Fig. 18. These results strongly support the CKM model: a selection of the parameter values obtained from the fit is shown in Table 2. From these results, constraints can be derived for New Physics (NP) contributions, beyond the CKM model.

For instance, the UTfit group performed a generalized fit [59] allowing for general NP loop contributions to $|\Delta F| = 2$ processes, and considering also the special case of extensions of the SM with "Minimal Flavour Violation". Taking into account also the first measurements of the B_s-$\overline{B}_s$ mass difference [26], they found strong constraints on NP contributions, probing energy scales up to ~5 TeV.

Table 2. Selected results from two global CKM fits: Wolfenstein CKM parameters $\bar{\rho}$, $\bar{\eta}$, and unitarity angles α, β and γ.

parameter	**CKMfitter** [32]	**UTfit** [41]
$\bar{\rho}$	$0.195^{+0.022}_{-0.055}$	0.164 ± 0.029
$\bar{\eta}$	$0.326^{+0.027}_{-0.015}$	0.340 ± 0.017
$\alpha(\phi_2)$ [°]	$99.0^{+4.0}_{-9.4}$	93.5 ± 4.6
$\beta(\phi_1)$ [°]	$22.03^{+0.72}_{-0.62}$	22.2 ± 0.9
$\gamma(\phi_3)$ [°]	$59.0^{+9.2}_{-3.7}$	64.1 ± 4.6

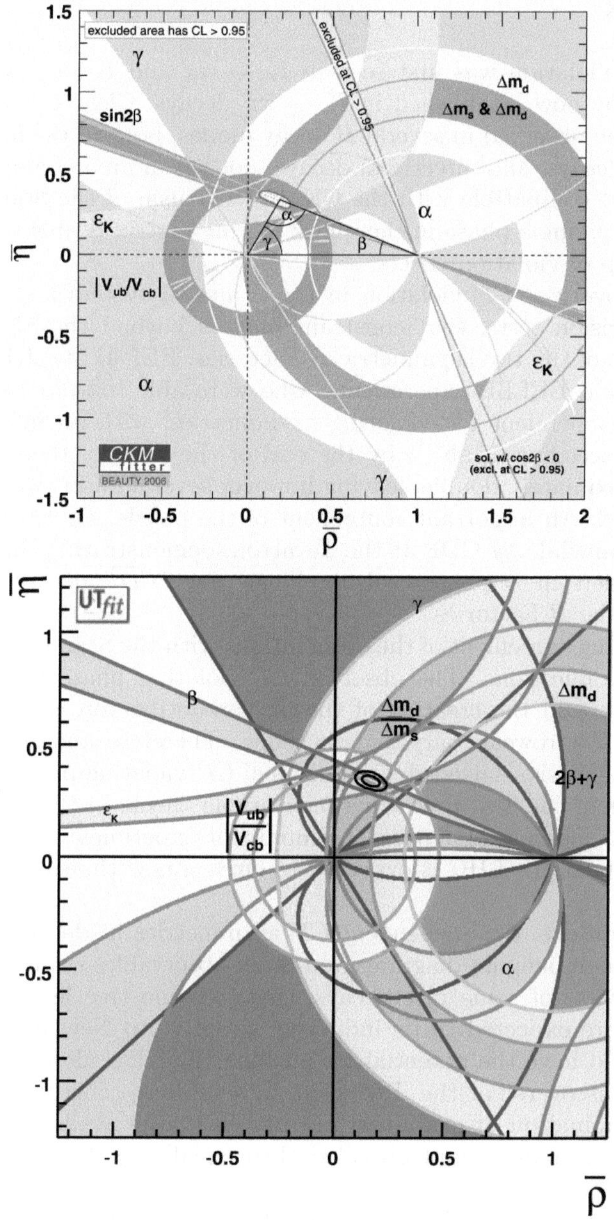

Fig. 18. (top) Confidence levels in the $(\bar{\rho}, \bar{\eta})$ plane obtained from the global CKM fit [32] of the CKMfitter group. The shaded areas indicate the regions of $\geq$5% CL; for $\sin 2\beta$ also the $\geq$32% constraint is shown. (bottom) Allowed regions for $(\bar{\rho}, \bar{\eta})$ from a fit performed by the UTfit group, with a Bayesian method. The closed contours at 68% and 95% probability are shown for the apex of the UT. The full lines correspond to 95% probability regions for the constraints, given by the measurements of $|V_{ub}|/|V_{cb}|$, ϵ_K, Δm_d, Δm_s, $\sin 2\beta$, γ, α and $\cos 2\beta$, respectively. The figure is taken from [41].

6 Outlook

Indirect *CP* violation was first seen in $K \to \pi\pi$ and $K \to \pi\ell\nu$, and direct violation is by now established in $K \to \pi\pi$ decays. Clear *CP* violation signals have been observed in several B decay modes, both in the interference of mixing and decays, and directly in decays, not yet in mixing alone. The measurements are compatible with the CKM mechanism of the Standard Model (SM), where a single phase in the quark mixing matrix is able to explain the entire pattern of violation effects.

The discovery of *CP* violation in the B meson system and the progress towards precision tests, that constrain physics beyond the SM, are recent achievements of the two asymmetric B Factories, PEP-II and KEK-B, and of the *BABAR* and BELLE experiments, who were able to systematically measure mixing-dependent *CP*-violating asymmetries, with an integrated total luminosity exceeding 1 ab^{-1}; by the end of the decade, their data sample should approximately double, leaving however several observables still statistically limited. An important component of the puzzle, B_s mixing, has been provided meanwhile by CDF at the Tevatron, demonstrating the potential of hadron colliders in B physics; also evidence for $D^0\overline{D}^0$ mixing was recently obtained at the B Factories.

The striking agreement of the observations with the Standard Model poses however new questions. The observed *CP*-violating phase is too small for the SM to explain the creation of the baryon-antibaryon asymmetry of the universe by electroweak baryogenesis alone. Therefore another mechanism, such as leptogenesis, is needed, or additional *CP*-violating phases should arise in extensions of the SM, producing measurable effects in the weak decays of heavy flavour mesons. Moreover, the imminent experimental investigation of the TeV scale at the LHC is expected to cross a new threshold and uncover new phenomena.

In this context, time-dependent *CP* asymmetries in decay channels proceeding through penguin diagrams and other observables such as the strongly suppressed rates of some rare decays, that have no tree level contributions in the SM, are expected to be indirectly sensitive to New Physics. Flavour physics should have the potential for pushing the NP scale up to 10-100 TeV or, if NP is discovered at the TeV scale, to be able to contribute significantly to the determination of couplings and phases, if the available luminosity is substantially increased, and provided theoretical uncertainties can be kept under control.

LHC experiments, and in particular the dedicated LHC-b spectrometer [60], will pursue the study of B physics with the unique capability of exploring *CP*-violating observables also in B_s decays. A very high luminosity B factory, taking advantage of the clean environment typical of an electron-positron machine, could have a complementary flavour physics reach, investigating those channels that are statistically limited or inaccessible at present, but that will not be easily measured in the hadron collider environment. Feasibility studies of such a machine are actively pursued at present [61], [62].

Finally, facilities are planned [63] also to study *CP* violation in very rare (a few $\times 10^{-11}$) and theoretically clean [64] but experimentally extremely challenging K meson decays: $K^+ \to \pi^+ \nu \overline{\nu}$ and $K_L^0 \to \pi^0 \nu \overline{\nu}$, with a goal of observing about a hundred events in each case.

Acknowledgements

I would like to thank our hosts for the warm hospitality in the beautiful environment of Akyaka, and to acknowledge the remarkable organization of the School: I specially enjoyed the opportunity of friendly and fruitful discussions with the other participants.

References

1. Branco G. C., Lavoura L. and Silva J. S., *CP Violation*, (Clarendon Press, Oxford) 1999
   Bigi I. I. and Sanda A. I., *CP Violation*, (Cambridge University Press, Cambridge, UK) 2000
   Kleinknecht K. *Uncovering CP violation: experimental clarification in the neutral K meson and B meson systems*, Springer Tracts in Modern Physics, Vol.195 (Springer Verlag) 2003
2. Kirkby D. and Nir Y., *CP Violation in Meson Decays* Ceccucci A., Ligeti Z., and Sakai Y., *The CKM Quark-Mixing Matrix* Schneider O., *B^0-antiB^0 mixing* Kwon Y. and Punzi G., *Production and Decay of b-Flavored Hadrons*, in [3]
3. Yao W.-M. et al. (Particle Data Group), J. Phys. **GB592**, 1 (2006), available on: http://pdg.lbl.gov/
4. Lanceri L., Experiments on *CP* violation with *B* mesons. In: *Proc. International School of Physics "Enrico Fermi"*, Course CLXIII, ed by M. Giorgi, I. Mannelli, A. I. Sanda, F. Costantini and M. S. Sozzi (IOS Press, Amsterdam and SIF, Bologna 2006) pp 247–308
5. Bloch P., *CPT Invariance Tests in Neutral Kaon Decay*, in [3]
6. Christenson J. H., Cronin J. W., Fitch V.L. and Turlay R., Phys. Rev. Lett. **13**, 138 (1964)
7. Sakharov, D. A., Sov. Phys. JETP Lett. **5**, 24 (1967)
8. Kobayashi M. and Maskawa T., Prog. Theor. Phys. **49**, 652 (1973)
9. Herb S. W. et al., Phys. Rev. Lett. **39**, 252 (1977)
10. Innes W. R. et al., Phys. Rev. Lett. **39**, 1240 (1977)
11. Ueno K. et al., Phys. Rev. Lett. **42**, 486 (1979)
12. Berger C. et al. (PLUTO Collaboration), Phys. Lett. B **76**, 243 (1978)
    Darden C. W. et al., Phys. Lett. **76**, 246 (1978)
    Bienlein J. K. et al., Phys. Lett. B **78**, 360 (1978)
    Darden C. W. et al., Phys. Lett. B **78**, 364 (1978)

13. Andrews D. et al. (CLEO Collaboration) Phys. Rev. Lett. **45**, 219 (1980)
Finocchiaro G. et al., Phys. Rev. Lett. **45**, 222 (1980)
Bebek C. et al. (CLEO Collaboration), Phys. Rev. Lett. **46**, 84 (1981)
14. Behrends S. et al. (CLEO Collaboration), Phys. Rev. Lett. **50**, 881 (1983)
15. Albajar C. et al. (UA1 Collaboration), Phys. Lett. B **186**, 247 (1987); Phys. Lett. B **197**, 565 (1987)
16. Albrecht H. et al. (ARGUS Collaboration), Phys. Lett. B **192**, 245 (1987)
17. Carter A. B. and Sanda A. I., Phys. Rev. Lett. **45**, 952 (1980); Phys. Rev. D **23**, 1567 (1981)
18. Bigi I. I. and Sanda A. I., Nucl. Phys. B **193**, 85 (1981); Nucl. Phys. B **281**, 41 (1987)
19. Oddone P., in: *Proceedings of the UCLA Workshop: Linear Collider $B\bar{B}$ Factory Conceptual Design*, edited by Stork D. (World Scientific, 1987), p. 243
20. PEP-II, *An Asymmetric B-Factory, Conceptual Design Report* SLAC-418, LBL-5379 (1993)
21. Kurokawa S. and Kikutani E., Nucl. Instr. Methods Phys. Res., Sect. A **499**, 1 (2003)
22. Aubert B. et al. (*BABAR* Collaboration), Phys. Rev. Lett. **89**, 201802 (2002)
Abe K. et al. (BELLE Collaboration), Phys. Rev. D **66**, 071102 (2002)
23. Aubert B. et al. (*BABAR* Collaboration), Phys. Rev. Lett. **93**, 131801 (2004)
24. Chao Y. et al. (BELLE Collaboration), Phys. Rev. Lett. **93**, 191802 (2004)
25. Weisskopf V. F. and Wigner E. P., Z. Phys. **63**, 54 (1930); Z. Phys. **65**, 18 (1930)
26. Abulencia A. et al., Phys. Rev. Lett. **97**, 242003 (2006)
27. For a recent review of results on D mixing, see [38]
28. Wolfenstein L. and Trippe T. G., *CP Violation in K_L^0 Decays*, in [3]
29. Wolfenstein L., Phys. Rev. Lett. **51**, 1945 (1983)
30. Buras A. J., Lautenbacher M. E., and Ostermaier G., Phys. Rev. D **50**, 3433 (1994)
31. Jarlskog C., Phys. Rev. Lett. **55**, 1039 (1985); Z. Phys. C **29**, 491 (1985)
32. Charles J. et al. (CKMfitter Group), Eur. Phys. Jour. C **41**, 1-131 (2005), arXiv:hep-ph/0406184, updated results and plots available at http://ckmfitter.in2p3.fr
33. Bander M., Silverman D. and Soni A., Phys. Rev. Lett. **43**, 242 (1979)
34. Laplace S., Ligeti Z., Nir Y. and Perez G., Phys. Rev. D **65**, 094040 (2002)
35. Aubert B. et al. (*BABAR* Collaboration), Nucl. Instr. Methods Phys. Res., Sect. A **479**, 1 (2002)
Abashian A. et al. (BELLE Collaboration), Nucl. Instr. Methods Phys. Res., Sect. A **479**, 117 (2002)
36. *BABAR* Collaboration, BABAR-CONF/00-01, SLAC-PUB-8450.
37. Aubert B. et al. (*BABAR* Collaboration), Phys. Rev. Lett. **95**, 151803 (2005); arXiv:hep-ex/0608035
38. Barberio E. et al. (Heavy Flavor Averaging Group), arXiv:hep-ex/0704.3575; updates available in http://www.slac.stanford.edu/xorg/hfag/
39. Alavi-Harati A. et al. (KTeV Collaboration), Phys. Rev. Lett. **83**, 22 (1999)
Fanti V. et al. (NA48 Collaboration), Phys. Lett. B **465**, 335 (1999)
Barr G. D. et al. (NA31 Collaboration), Phys. Lett. B **317**, 233 (1999)
40. Angelopoulos A. et al., Eur. Phys. Jour. C **18**, 41 (2000)

41. Bona M. et al. (UTfit Collaboration), Journal of High Energy Physics 028, 0507 (2005); arXiv:hep-ph/0605213; arXiv:hep-ph/0606167; arXiv:hep-ph/0701204 updates available in http://www.utfit.org/
42. Aubert B. et al. (*BABAR* Collaboration), arXiv:hep-ex/0703021v1
43. Chen K.-F. et al. (BELLE Collaboration), Phys. Rev. Lett. **98**, 031802 (2007)
44. Gronau M. and London D., Phys. Rev. Lett. **65**, 3381 (1990)
45. Ishino H. et al. (BELLE Collaboration), arXiv:hep-ex/0608035v3
46. Aubert B. et al. (*BABAR* Collaboration), arXiv:hp-ex/0612021
47. Quinn H. R. and Snyder A. E., Phys. Rev. D **48**, 2139 (1993)
48. Lipkin H. J., Nir Y., Quinn H. R., and Snyder A., Phys. Rev. D **44**, 1454 (1991).
49. Aubert B. et al. (*BABAR* Collaboration), arXiv:hep-ex/0703008
50. Kusaka A., Wang C. C., Ishino H., et al., (BELLE Collaboration) arXiv:hep-ex0701015
51. See, for example, Neubert M. and Stech B., in Heavy Flavors, 2nd Edition, edited by Buras A. J. and Lindner M., World Scientific, Singapore, 1997; Gronau M., Phys. Lett. B **557** 198 (2003)
52. Gronau M. and Wyler D., Phys. Lett. B **265**, 172 (1991)
Gronau M. and London D., Phys. Lett. B **253**, 483 (1991)
53. Atwood A., Dunietz I., and Soni A., Phys. Rev. Lett. **78**, 3257 (1997)
54. Giri A., Grossman Y., Soffer A., and Zupan J., Phys. Rev. D **68**, 054018 (2003)
55. Dunietz I., Phys. Lett. B **427**, 179 (1998)
56. Aubert B. et al. (*BABAR* Collaboration), Phys. Rev. D **71**, 112003 (2005); Phys. Rev. D **73**, 111101 (2006)
57. Ronga F. J., et al (BELLE Collaboration), Phys. Rev. D **73**, 092003 (2006)
58. Battaglia M. et al., arXiv:hep-ph/0304132; Schubert K. R., Eur. Phys. Jour. **A 18**, 147 (2003); Abele H. et al., Eur. Phys. Jour. C **33**, 1 (2004); Ali A., arXiv:hep-ph/0312303; Eigen G., Dubois-Felsmann G. P., Hitlin D. G., and Porter F. C., Eur. Phys. Jour. C **33**, S644 (2004); Ratcliffe P. G., arXiv:hep-ph/0402063; Stocchi A., arXiv:hep-ph/0405038; Charles J. et al., arXiv:hep-ph/0406184; Bona M. et al., arXiv:hep-ph/0408079; Heavy Flavor Averaging Group, arXiv:hep-ex/0505100; Yao W.-M. et al. (Particle Data Group), J. Phys. **GB592**, 1 (2006)
59. Bona M. et al, UTfit Collaboration, arXiv:hep-ex/0605213v3
60. LHCb Technical Proposal, CERN-LHCC-98-004, LHCC/P4, 20 February 1998; LHCb Technical Design Report-9, CERN-LHCC-2003-064/G-063
61. *SuperB, A High-Luminosity Heavy Flavour Factory*, Conceptual Design Report, INFN/AE-07/2, SLAC-R-856, LAL 07-15, March, 2007
62. SuperKEKB Letter of Intent, KEK Report 04-4, August, 2004
63. *Proposal to Measure the Rare Decay $K^+ \to \pi^+ \nu \overline{\nu}$ at the CERN SPS*, CERN-SPSC-2005-013, SPSC-P-326, 11.6.2005
Komatsubara T. K., *Future Kaon Program at KEK/J-PARC*, arXiv:hep-ex/0409017v1
64. Buras A. J., Ewerth T., and Jager S., Nucl. Phys. B **714**, 103 (2005)

Worldline Formalism and It's Application to AdS/CFT Correspondence

Sh. Mamedov

Institute for Physical Problems, Baku State University, Z. Khalilov 23, Baku, AZ/1148 AzerbaijanSh_-Mamedov@bsu.az

LECTURE 1

The Worldline Formalism

One of the interesting features of string theory is that, it reduces to quantum field theory in the infinite string tension limit. Consequently, in this limit the string scattering amplitudes should reduce to corresponding field theory amplitudes. First realization of this idea was the derivation of the one-loop β-function coefficient for the pure Yang-Mills theory from the partition function of an open string propagating in a Yang-Mills background [10]. Another investigation of the infinite string tension limit was done by Bern and Kosower [11] with the aim of finding useful results to be used for QCD calculations. The idea was to derive a parameter integral expression for gluon - gluon scattering amplitude from the string scattering amplitude in a superstring theory. In the infinite string tension limit all massive modes are suppressed and massless string modes survive, and one obtains a parameter integral representation of string scattering amplitude. This corresponds to the gauge boson scattering amplitude in some parameter (Schwinger parameter) integral representation in the field theory. The rules obtained in this approach for the calculation of field theory amplitudes are named as Bern-Kosower rules. In [1], Bern-Kosower rules have been derived in the field theory framework, as a continuation of this idea, without applying string theory. So there appeared a new formalism named worldline formalism. This formalism was applied to a number of field theory [2], and gravity problems [12]. A good pedagogical introduction to worldline formalism was given in [3]. In this lecture we intend to give an introduction to worldline formalism based on papers [1, 2].

1 Scalar Particles

We will first consider an abelian theory and construct the worldline formalism for the scalar particles. In order to find one-loop effective action for scalar particles we should take the logarithm of the partition function

$$\Gamma = \log Z = \log \left[\int D\phi^\dagger D\phi \exp\left(i \int d^4x \mathcal{L} \right) \right] \tag{1}$$

$$\mathcal{L} = -\left(\partial\phi^\dagger\right) \cdot (\partial\phi) - m^2\phi^\dagger\phi.$$

As usual we throw away the surface term and write exponent in (1) in a suitable form for the Gaussian integration:

$$\int d^4x \left(\partial_\mu\phi^\dagger\right) \cdot (\partial^\mu\phi) = \int d^4x \partial_\mu \left(\phi^\dagger \partial^\mu \phi\right) - \int d^4x \phi^\dagger \left(\partial^\mu\partial_\mu\right)\phi = -\int d^4x \phi^\dagger \left(\partial^\mu\partial_\mu\right)\phi$$

$$\int d\phi \exp\left\{ -\frac{1}{2}\phi^\dagger A \phi \right\} = (\det A)^{-1/2} \tag{2}$$

In our case the infinite dimensional matrix A has a form:

$$A = -i\left(\partial^2 - m^2\right). \tag{3}$$

and scalar product means

$$\left(\phi^\dagger, \phi\right) = \int d^4x \left(\phi^\dagger\phi\right); \quad \left(\phi^\dagger, A\phi\right) = \int d^4x \left(\phi^\dagger(x) A\phi(x)\right)$$

The operator A is diagonal in the Hilbert space of eigenvectors and its determinant becomes a product of eigenvalues:

$$\det A = a_1 a_2 \cdots a_n = \prod_{i=1}^{n} a_i \tag{4}$$

Thus, having used formula (2) twice in the effective action (1) it has got following form:

$$\log Z = \log\left[\det\left(-\partial^2 + m^2\right)\right]^{-1} = -\log\left[\det\left(-\partial^2 + m^2\right)\right]. \tag{5}$$

Note that, the operator (3) has a zero mode and in order to avoid the corresponding singularity we have to work in subspace of Hilbert space, obtained by removing the eigenvector corresponding to this mode. Using formula $\log\det A = Tr \log A$ and remembering the definition of functional trace formula:

$$Tr \log A = \int d^4x \; \langle x \left|\log A\right| x \rangle \tag{6}$$

we can convert functional determinant in (5) into functional trace

$$\log Z = -Tr \log\left(-\partial^2 + m^2\right) = Tr \log\left[1/\left(-\partial^2 + m^2\right)\right]$$
$$= \int d^4x \left\langle x \left| \log\left[1/\left(-\partial^2 + m^2\right)\right] \right| x \right\rangle \tag{7}$$

We next go to momentum space

$$\Gamma = \log Z = \int d^4x \int d^4p \int d^4p' \left\langle x \mid p \right\rangle \left\langle p \left| \log\left[1/\left(-\partial^2 + m^2\right)\right] \right| p' \right\rangle \left\langle p' \mid x \right\rangle$$

and use the following integral representation for the logarithm

$$\log A = \int_0^\infty \frac{dT}{T} \exp\left[iTA\right].$$

Then the effective action will get following simple form:

$$\Gamma = \int d^4p \int d^4p' \int d^4x \frac{e^{ipx}}{(2\pi)^2} \left\langle p \left| \int_0^\infty \frac{dT}{T} \exp\left[iT\left(-\partial^2 + m^2\right)\right] \right| p' \right\rangle \frac{e^{-ip'x}}{(2\pi)^2}$$
$$= \int d^4p \int d^4p' \int d^4x \frac{e^{i(p-p')x}}{(2\pi)^4} \int_0^\infty \frac{dT}{T} \left\langle p \left| \exp\left[iT\left(p'^2 + m^2\right)\right] \right| p' \right\rangle$$
$$= \int d^4p \int_0^\infty \frac{dT}{T} \left\langle p \left| \exp\left[iT\left(p^2 + m^2\right)\right] \right| p \right\rangle$$
$$= \int \frac{d^4p}{(2\pi)^4} \int_0^\infty \frac{dT}{T} \exp\left[iT\left(p^2 + m^2\right)\right].$$

Here we have used $\langle x_i \mid p_i \rangle = e^{ip_ix_i}/\left(2\pi\right)^2$ and

$$\left\langle p_i \left| \exp\left[-\tau\left(-\partial^2 + m^2\right)\right] \right| p_i' \right\rangle = \delta\left(p_i - p_i'\right) e^{-\tau\left(p^2(\tau_i) + m^2\right)}.$$

After Wick rotation of contour, the effective action becomes:

$$\Gamma = \int_0^\infty \frac{dT}{T} \int \frac{d^4p}{(2\pi)^4} \exp \frac{\mathcal{E}}{2} \left[-T\left(p^2 + m^2\right)\right]. \tag{8}$$

Here we have inserted constant $\mathcal{E}$, which will be set equal to 2 after some transformations. Now divide T into N equal parts $T = \sum_{i=1}^{N+1} \tau$ and suppose that coordinate and momentum are functions of τ $x(\tau)$, $p(\tau)$. We shall have

N values of these functions $x_i = x(\tau_i)$, $p_i = p(\tau_i)$ and according to the trace formula (6) it is reasonable to choose $x_1 = x$, $x_{N+1} = x$. Then effective action (7) will be transformed into integrals over the momenta and coordinates at these points:

$$\Gamma = \int d^4x \left\langle x \left| \log\left[1/\left(-\partial^2 + m^2\right)\right]\right| x\right\rangle$$

$$= \int_0^\infty \frac{dT}{T} \int d^4x \left\langle x \left| \exp\left[-iT\left(-\partial^2 + m^2\right)\right]\right| x\right\rangle$$

$$= \int_0^\infty \frac{dT}{T} \int d^4x_1 \cdots \int d^4x_N \left\langle x_1 \left| \exp\left[-\tau\left(-\partial^2 + m^2\right)\right]\right| x_2\right\rangle \cdot$$

$$\left\langle x_2 \left| \exp\left[-\tau\left(-\partial^2 + m^2\right)\right]\right| x_3\right\rangle \cdots \cdot \left\langle x_N \left| \exp\left[-\tau\left(-\partial^2 + m^2\right)\right]\right| x_{N+1}\right\rangle$$

$$= \int_0^\infty \frac{dT}{T} \int d^4x_1 \cdots \int d^4x_N \int d^4p_1 \cdots \int d^4p_{N+1} \int d^4p'_1 \cdots \int d^4p'_{N+1} \cdot$$

$$\cdot \left\langle x_1 \mid p_1\right\rangle \left\langle p_1 \left| \exp\left[-\tau\left(-\partial^2 + m^2\right)\right]\right| p'_1\right\rangle \left\langle p'_1 \mid x_2\right\rangle$$

$$\cdot \left\langle x_2 \mid p_2\right\rangle \left\langle p_2 \left| \exp\left[-\tau\left(-\partial^2 + m^2\right)\right]\right| p'_2\right\rangle \left\langle p'_2 \mid x_3\right\rangle \cdots$$

$$\cdots \cdot \left\langle x_N \mid p_{N+1}\right\rangle \left\langle p_{N+1} \left| \exp\left[-\tau\left(-\partial^2 + m^2\right)\right]\right| p'_{N+1}\right\rangle \left\langle p'_{N+1} \mid x_{N+1}\right\rangle$$

$$= \int_0^\infty \frac{dT}{T} \int d^4x_1 \cdots \int d^4x_N \int d^4p_1 \cdots \int d^4p_{N+1} \int d^4p'_1 \cdots \int d^4p'_{N+1} \cdot$$

$$\frac{e^{ip_1x_1}}{(2\pi)^2} \delta\left(p_1 - p'_1\right) e^{-\tau\left(p^2(\tau_1)+m^2\right)} \frac{e^{-ip'_1x_2}}{(2\pi)^2} \frac{e^{ip_2x_2}}{(2\pi)^2} \delta\left(p_2 - p'_2\right) e^{-\tau\left(p^2(\tau_2)+m^2\right)} \frac{e^{-ip'_2x_3}}{(2\pi)^2}$$

$$\cdots \cdot \frac{e^{ip_Nx_N}}{(2\pi)^2} \delta\left(p_{N+1} - p'_{N+1}\right) e^{-\tau\left(p^2(\tau_N)+m^2\right)} \frac{e^{-ip'_{N+1}x_{N+1}}}{(2\pi)^2}$$

Taking integrals over the δ functions we obtain:

$$\Gamma = \int_0^\infty \frac{dT}{T} \int d^4x_1 \cdots \int d^4x_N \int d^4p_1 \cdots$$

$$\cdots \int d^4p_{N+1} \frac{e^{-ip_1(x_2-x_1)}}{(2\pi)^2} \exp\left\{-\tau\left(p^2(\tau_1) + m^2\right)\right\} \cdots \cdot$$

$$\cdots \frac{e^{-ip_{N+1}(x_{N+1}-x_N)}}{(2\pi)^2} \exp\left\{-\tau\left(p^2(\tau_N) + m^2\right)\right\}$$

For large N we can write

$$ip_N\left(x_{N+1}-x_N\right)=ip_N\Delta x_N=ip_N\,\dot{x}\left(\tau_N\right)d\tau,$$

where dot denotes derivative over the τ, and in the $N\to\infty$ limit the effective action becomes following path integrals of x and p:

$$\lim_{N\to\infty}\log Z=\int_0^\infty\frac{dT}{T}\int\limits_{x(\tau)=x(0)}Dx\int Dp\,\exp\left\{-\int_0^T d\tau\left(p^2\left(\tau\right)+m^2\right)\right\}\cdot\exp\left\{-i\int_0^T d\tau\;p\left(\tau\right)\dot{x}\left(\tau\right)\right\}.$$

Shifting the momentum as,

$$p'\left(\tau\right)\to-p\left(\tau\right)+\frac{i}{\mathcal{E}}\,\dot{x}\left(\tau\right)$$

we can factorize this expression into a product of two independent path integrals:

$$\log Z=\int_0^\infty\frac{dT}{T}\int\limits_{x(\tau)=x(0)}Dx\int Dp\,\exp\left\{-\frac{1}{2}\mathcal{E}\int_0^T d\tau p^2\left(\tau\right)\right\}\exp\left\{-\int_0^T d\tau\left(\frac{1}{2\mathcal{E}}\,\dot{x}^2\left(\tau\right)+\frac{\mathcal{E}}{2}m^2\right)\right\}$$

$$=\int_0^\infty\frac{dT}{T}\mathcal{N}\int\limits_{x(\tau)=x(0)}Dx\exp\left\{-\int_0^T d\tau\left(\frac{1}{2\mathcal{E}}\,\dot{x}^2\left(\tau\right)+\frac{\mathcal{E}}{2}m^2\right)\right\}.\qquad(9)$$

Here $\mathcal{N}$ denotes

$$\mathcal{N}=\int Dp\;\exp\left\{-\frac{1}{2}\mathcal{E}\int_0^T d\tau p^2\left(\tau\right)\right\}$$

We have obtained two different expressions (8) and (9) for the effective action, which, of course, are equal to each-other:

$$\mathcal{N}\int\limits_{x(\tau)=x(0)}Dx\exp\left\{-\int_0^T d\tau\frac{1}{2\mathcal{E}}\,\dot{x}^2\left(\tau\right)\right\}=\int\frac{d^Dp}{\left(2\pi\right)^D}\exp\left[-\frac{\mathcal{E}}{2}Tp^2\right]=\left(2\pi\mathcal{E}T\right)^{-D/2}\qquad(10)$$

Let us consider scalar field theory interacting with an external field. The Lagrangian of this theory can be obtained from the free theory by the momentum shift:

$$D_\mu=\partial_\mu-igA_\mu.\qquad(11)$$

After repeating the transformations above, for the effective action with the new Lagrangian (11) we will get the following expression:

$$\Gamma\left[A\right]=-\log Z=-\log\left[\det\left(-D^2\right)\right]=\int_0^\infty\frac{dT}{T}\int\frac{d^4p}{\left(2\pi\right)^4}\exp\left[-\frac{\mathcal{E}}{2}T\left(p+gA\left(x\right)\right)^2\right].$$

The alternative transformation, similar to the one done for the free theory, will reduce effective action of interacting theory into following path integrals:

$$\Gamma[A]=\int_0^\infty \frac{dT}{T} \int_{x(\tau)=x(0)} Dx \int Dp \exp\left\{i\int_0^T d\tau p(\tau)\,\dot{x}(\tau)\right\} \exp\left\{-\frac{\mathcal{E}}{2}\int_0^T d\tau\,[p(\tau)+gA(x(\tau))]^2\right\}.$$

Now we shift the momentum as,

$$p(\tau) = p'(\tau) + \frac{i}{\mathcal{E}}\,\dot{x}(\tau) - gAx(\tau)$$

This allows us to separate the momentum path integral in the effective action (10):

$$\begin{aligned}
\Gamma[A] &= \int_0^\infty \frac{dT}{T} \int Dp\ \exp\left\{-\mathcal{E}\int_0^T d\tau p^2(\tau)\right\} \\
&\times \int_{x(\tau)=x(0)} Dx \exp\left\{-\int_0^T d\tau \left[\frac{1}{2\mathcal{E}}\,\dot{x}^2(\tau) + igA(x(\tau))\,\dot{x}(\tau)\right]\right\} \\
&= \int_0^\infty \frac{dT}{T}\mathcal{N} \int_{x(\tau)=x(0)} Dx\ \exp\left\{-\int_0^T d\tau \left[\frac{1}{2\mathcal{E}}\,\dot{x}^2(\tau) + igA(x(\tau))\,\dot{x}(\tau)\right]\right\}.
\end{aligned}$$

Thus we have only one path integral in the effective action expression. Continuing $\Gamma[A]$ into Minkowski space and redefining $\mathcal{E} \to -\mathcal{E}$ we get:

$$\begin{aligned}
\Gamma[A] &= \int_0^\infty \frac{dT}{T}\mathcal{N} \int_{x(\tau)=x(0)} Dx\ \exp\left\{-\int_0^T d\tau \left[\frac{1}{2\mathcal{E}}\,\dot{x}^2(\tau) - igA(x(\tau))\,\dot{x}(\tau)\right]\right\} \\
&= \int_0^\infty \frac{dT}{T}\mathcal{N} \int_{x(\tau)=x(0)} Dx\ \exp\left\{-\int_0^T d\tau \left[\frac{1}{2\mathcal{E}}\,\dot{x}^2(\tau)\right]\right\} \exp\left(ig\oint dx\cdot A(x)\right)
\end{aligned} \tag{12}$$

In non-abelian case, the trace in (1) will include the trace over the color indices as well, and our final expression for the effective action (12) will include this trace:

$$\Gamma[A] = \int_0^\infty \frac{dT}{T}\mathcal{N} \int_{x(\tau)=x(0)} Dx\ Tr\ \exp\left\{-\int_0^T d\tau \left[\frac{1}{2\mathcal{E}}\,\dot{x}^2(\tau) + igA(x(\tau))\,\dot{x}(\tau)\right]\right\}$$

Now decompose (12) in powers of g

$$\Gamma[A] = \sum_{N=1}^{\infty} \int_0^\infty \frac{dT}{T}\mathcal{N} \int_{x(\tau)=x(0)} Dx\ Tr\ \exp\left\{-\frac{1}{2\mathcal{E}}\int_0^T d\tau\,\dot{x}^2(\tau)\right\}$$

$$\times \frac{(ig)^N}{N!} \prod_{i=1}^{N} \int_0^T dt_i \; A\left[x\left(t_i\right)\right] \dot{x}\left(t_i\right) = \sum_{N=1}^{\infty} \Gamma_N\left[A\right]$$

and choose A^μ as a sum of plane waves with polarization vectors ϵ_j^μ

$$A^\mu\left(x\right) = \sum_{j=1}^{N} T^{a_j} \epsilon_j^\mu \; e^{ik_j \cdot x}.$$

Then each term in this decomposition has the form:

$$\Gamma_N\left(k_1, ...k_N\right) = \left(ig\right)^N \int_0^\infty \frac{dT}{T} \mathcal{N} \int_{x(\tau)=x(0)} Dx \; \exp\left\{-\frac{1}{2\mathcal{E}} \int_0^T d\tau \; \dot{x}^2\left(\tau\right)\right\}$$

$$\times Tr\left(T^{a_N}......T^{a_1}\right) \prod_{j=1}^{N} \int_0^{t_{j+1}} dt_j \; \epsilon_j \cdot \dot{x}\left(t_i\right) \; e^{ik_j \cdot x(t_j)} \tag{13}$$

Here we have a factor containing the polarization vector; this stops us from making further transformations of exponents. For simplicity we first choose $\epsilon_j \cdot \dot{x}\left(t_i\right) = 1$, and continue our transformations:

$$\mathcal{N} \int_{x(\tau)=x(0)} Dx \quad \exp\left\{-\frac{1}{2\mathcal{E}} \int_0^T d\tau \; \dot{x}^2\left(\tau\right) + ik_j \cdot x\left(t_i\right)\right\}$$

$$= \mathcal{N} \int_{x(\tau)=x(0)} Dx \quad \exp\left\{-\int_0^T d\tau \left[-\frac{1}{2\mathcal{E}} x\left(\tau\right) \frac{d^2}{d\tau^2} x\left(\tau\right) - ik_j \cdot x\left(\tau\right) \delta\left(\tau - t_j\right)\right]\right\}$$

$$= \mathcal{N} \int_{x(\tau)=x(0)} Dx \quad \exp\left\{-\int_0^T d\tau \left[-\frac{1}{2\mathcal{E}} x\left(\tau\right) \frac{d^2}{d\tau^2} x\left(\tau\right) - J^\mu\left(\tau\right) \cdot x\left(\tau\right)\right]\right\}$$

$$= \mathcal{N} \int_{x(\tau)=x(0)} Dx \quad \exp\left\{-\frac{1}{2\mathcal{E}} \int_0^T d\tau \; \dot{x}^2\left(\tau\right)\right\}$$

$$\times \exp\left\{-\frac{1}{2} \int_0^T d\tau \int_0^T d\tau' J^\mu\left(\tau\right) G_B\left(\tau, \tau'\right) J_\mu\left(\tau'\right)\right\}$$

$$= \left(2\pi \mathcal{E} T\right)^{-2} \cdot \exp\left\{-\frac{1}{2} \int_0^T d\tau \int_0^T d\tau' J^\mu\left(\tau\right) G_B\left(\tau, \tau'\right) J_\mu\left(\tau'\right)\right\}$$

Here we have denoted $J^{\mu}(\tau)=\sum_{j=1}^{N} ik_j^{\mu}\ \delta(\tau-t_j)$ and introduced Green's function for the equation

$$\frac{1}{\mathcal{E}}\partial_t^2 G_B(t,t')=\delta(t-t')\rightarrow G_B(t,t')=\frac{1}{2}\mathcal{E}\left|t-t'\right|+A+Bt$$

which is called worldline bosonic Green's function. But this solution for the equation on the real line has no singularity at $t\rightarrow t'$. So we redefine G_B for the equation by shifting at $-\frac{1}{T}$

$$\frac{1}{\mathcal{E}}\partial_t^2 G_B(t,t')=\delta(t-t')-\frac{1}{T},$$

which under periodicity condition has following solution:

$$G_B(t,t')=\frac{1}{2}\mathcal{E}\left(\left|t-t'\right|-\left(t-t'\right)^2\right)+A.$$

Furthermore for G_B we use notations:

$$G_B(t_i,t_j)\equiv G_B^{ij},\ \dot{G}_B^{ij}\equiv\partial_{t_i}G_B^{ij},\ \ddot{G}_B^{ij}\equiv\partial_{t_i}^2 G_B^{ij}.$$

For the choice $\epsilon_j\cdot\dot{x}(t_i)=1$ the one loop N-point amplitude will get a form:

$$\Gamma_N(k_1,.....k_N)=\frac{(ig)^N}{(4\pi)^2(\mathcal{E}/2)^2}Tr\left(T^{a_N}......T^{a_1}\right)\int_0^{\infty}\frac{dT}{T^3}$$

$$\times\prod_{j=1}^{N}\int_0^{t_{j+1}}dt_j\exp\left\{-\frac{1}{2}\int_0^T d\tau\int_0^T d\tau' J^{\mu}(\tau)G_B(\tau,\tau')J_{\mu}(\tau')\right\}$$

$$=\frac{(ig)^N}{(4\pi)^2(\mathcal{E}/2)^2}Tr\left(T^{a_N}......T^{a_1}\right)\int_0^{\infty}\frac{dT}{T^3}\prod_{j=1}^{N}\int_0^{t_{j+1}}dt_j\cdot\exp\left[\frac{1}{2}\sum_{i,j=1}^{N}k_ik_jG_B(t_i,t_j)\right]$$

Now we can put back the polarization vectors and transform (13). That is, we can now handle exponents. In order to write ϵ_j in the exponent let us apply the following formula:

$$\dot{x}(t_j)A_i^{\mu}(x(t_i))=T^{a_j}\exp\left[\epsilon_j\cdot\partial_{t_j}x(t_j)+ik_j\cdot x(t_j)\right]|_{linear\ in\ \epsilon_j}$$

So, current J^{μ} will get shifted due to ϵ_j

$$J^{\mu}(\tau)=\sum_{j=1}^{N}\delta(\tau-t_j)\left[\epsilon_j^{\mu}\partial_{t_j}+ik_j^{\mu}\right]$$

and the exponent in the N-point amplitude becomes

$$\exp\left\{-\frac{1}{2}\int_0^T d\tau\int_0^T d\tau' J^{\mu}(\tau)G_B(\tau,\tau')J_{\mu}(\tau')\right\}=\exp\left[-\frac{1}{2}\int_0^T d\tau\int_0^T d\tau'\right.$$

$$\times \sum_{i,j=1}^{N} \delta(\tau - t_j)\,\delta(\tau' - t_i)\left[\epsilon_i^{\mu}\partial_{t_i} + ik_i^{\mu}\right]\left[\epsilon_{\mu}^{j}\partial_{t_j} + ik_{\mu}^{j}\right] G_B(t_i,t_j)\Bigg]$$

$$= \exp\left\{ -\frac{1}{2}\int_0^T d\tau \int_0^T d\tau' \left[\sum_{i,j=1}^{N} \delta(\tau - t_j)\,\delta(\tau' - t_i)\,\epsilon_i^{\mu}\epsilon_{\mu}^{j}\partial_{t_i}\partial_{t_j} + i\epsilon_i^{\mu}k_{\mu}^{j}\partial_{t_i} \right.\right.$$

$$\left.\left. + i\epsilon_{\mu}^{j}k_i^{\mu}\partial_{t_j} - k_i^{\mu}k_{\mu}^{j}\right] G_B(t_i,t_j)\right\}$$

It is useful to replace $t_i \to u_i T$, where u_i is dimensionless. $\Gamma_N(k_1,.....k_N)$ in terms of these variables attains a form ready for application to calculations of one loop N tail irreducible diagrams:

$$\Gamma_N(k_1,.....k_N) = \frac{\left(ig\mu^{\epsilon/2}\right)^N}{(4\pi)^{2-\epsilon/2}} Tr\left(T^{a_N}......T^{a_1}\right)\int_0^{\infty}\frac{dT}{T^{3-N}}$$

$$\times \int_0^1 du_{N-1}\int_0^{u_{N-1}} du_{N-2}\cdots\int_0^{u_2} du_1 \exp\left[\sum_{i<j=1}^{N} k_i k_j G_B^{ji}\right]$$

$$\times \exp\left[\sum_{i<j=1}^{N}\left(-i\left(k_i\epsilon_j - \epsilon_i k_j\right)\dot{G}_B^{ji} + \epsilon_i\epsilon_j \ddot{G}_B^{ji}\right)\right]\Big|_{linear\ in\ each\ \epsilon} \tag{14}$$

In terms of the u_i variables, worldline Green's function becomes:

$$G_B(t,t') = T\left(|u-u'| - (u-u')^2\right),\quad \partial_t G_B(t,t') = sign(u-u') - 2(u-u')$$

$$\partial_t^2 G_B(t,t') = \frac{2}{T}\left(\delta(u-u') - 1\right)$$

In order to check formula (14) we can set $N = 2$. This gives us two-point correlator

$$\Gamma_2(k_1,k_2) = \frac{\left(ig\mu^{\epsilon/2}\right)^2}{(4\pi)^{2-\epsilon/2}} Tr\left(T^aT^b\right)\int_0^{\infty}\frac{dT}{T^{1-\epsilon/2}}\int_0^1 du\; e^{k_1\cdot k_2 G_B(1-u)}$$

$$\times\left\{k_2\cdot\epsilon_1 k_1\cdot\epsilon_2\left[\dot{G}_B(1-u)\right]^2 + \epsilon_1\cdot\epsilon_2\,\ddot{G}_B(1-u)\right\}, \tag{15}$$

which is known in QED as the polarization operator and has the expression:

$$\Pi(k_1,k_2) = \frac{-\left(g\mu^{\epsilon/2}\right)^2}{(4\pi)^{2-\epsilon/2}} Tr\left(T^aT^b\right)\int_0^{\infty}\frac{dT}{T^{1-\epsilon/2}}\int_0^1 da\; e^{Tk_1\cdot k_2\left(a-a^2\right)}$$

$$\times\left\{\frac{2}{T}\epsilon_1\cdot\epsilon_2\left(\delta(a) - 1\right) + (1-2a)^2 k_2\cdot\epsilon_1 k_1\cdot\epsilon_2\right\}$$

It can be easily checked that the expressions of the correlator obtained by the worldline formalism and direct calculation in the Schwinger proper time respectively are same. This confirm the correctness of the worldline formalism.

2 Spinor Particle

Let us consider now a spinning particle in the loop. The action for such a particle in a background field A_μ is

$$S = \int d^4x\, \overline{\chi} \left(i\widehat{D} - m \right) \chi \quad ; \quad D_\mu = \partial_\mu - igA_\mu$$

The one-loop effective action is obtained by using the formula

$$\int d\overline{\alpha} d\alpha\; e^{-\overline{\alpha} A \alpha} = \det A = e^{Tr \log A}$$

$$= \frac{1}{2} \log \left[\det \left(\widehat{D}^2 + m^2 \right) \right] = \frac{1}{2} \log \left[\det \left(D^2 - \frac{1}{4} igF_{\mu\nu} \left[\gamma_\mu, \gamma_\nu \right] + m^2 \right) \right]$$

$$= \frac{1}{2} Tr \log \left(D^2 - \frac{1}{4} igF_{\mu\nu} \left[\gamma_\mu, \gamma_\nu \right] + m^2 \right)$$

Thus for the spinor particle the effective action can be expressed as:

$$\Gamma \left[A \right] = \frac{1}{2} Tr \log \left(D^2 - \frac{1}{4} igF_{\mu\nu} \left[\gamma_\mu, \gamma_\nu \right] + m^2 \right) \tag{16}$$

Let us note that here we have different sign from the effective action for the scalar loop.

2.1 Coherent State Formalism

It is seen from (16) that the effective action contains trace over the γ matrices. These matrices create some difficulties in converting the functional trace in (16) into the path integrals. But problem with γ matrices is not new, and there exists a formalism, which converts trace over these matrices into path integrals over Grassmann variables [4]. This formalism is called coherent state formalism, and is successfully applied for the constructing worldline path integrals [2].

Let us form operators $a_i^\pm$ from γ matrices

$$a_1^\pm = \frac{1}{2} \left(\gamma_1 \pm i\gamma_3 \right), \quad a_2^\pm = \frac{1}{2} \left(\gamma_2 \pm i\gamma_4 \right)$$

According to anticommutation relation between γ matrices these operators obey following anticommutation relations:

$$\left\{ a_r^+, a_s^- \right\} = \delta_{rs}, \quad \left\{ a_r^+, a_s^+ \right\} = 0, \quad \left\{ a_r^-, a_s^- \right\} = 0.$$

Thus, they can be considered as the creation and annihilation operators in some space, for which we can define vacuum states as:

$$a_s^- |0\rangle = 0, \quad \langle 0| a_s^+ = 0$$

Let us introduce Grassmann variables η_r and $\overline{\eta}_r$ ($r = 1,2$) $\eta_r \overline{\eta}_r = -\overline{\eta}_r \eta_r$ and define new states by means of these variables and $a^{\pm}$ operators in the following way

$$\langle \eta| \equiv i \langle 0| \left(\eta_1 - a_1^-\right) \left(\eta_2 - a_2^-\right) \qquad |\eta\rangle \equiv \exp\left(-\eta_1 a_1^+ - \eta_2 a_2^+\right) |0\rangle$$

$$\langle \overline{\eta}| \equiv \langle 0| \exp\left(-a_1^- \overline{\eta}_1 - a_2^- \overline{\eta}_2\right) \qquad |\overline{\eta}\rangle \equiv i\left(\overline{\eta}_1 - a_1^+\right)\left(\overline{\eta}_2 - a_2^+\right)|0\rangle \quad (17)$$

It can be checked that action of $a^{\pm}$ operators on these states are:

$$\langle \eta| a_r^- = \langle \eta| \eta_r \qquad a_r^- |\eta\rangle = \eta_r |\eta\rangle ,$$

$$\langle \overline{\eta}| a_r^+ = \langle \overline{\eta}| \eta_r \qquad a_r^+ |\overline{\eta}\rangle = \overline{\eta}_r |\overline{\eta}\rangle \quad (18)$$

i.e. these states are eigenstates of $a^{\pm}$ operators and so can be considered as the coherent states in the above defined space. Products of these states are equal to:

$$\langle \eta| |\overline{\eta}\rangle = \exp\left(\eta_1 \overline{\eta}_1 + \eta_2 \overline{\eta}_2\right)$$

$$\langle \overline{\eta}| |\eta\rangle = \exp\left(\overline{\eta}_1 \eta_1 + \overline{\eta}_2 \eta_2\right)$$

and has a resemblance to coherent states in an ordinary Quantum Mechanics. Integrals over the Grassmann variables are defined according to the rules:

$$\int \eta_i d\eta_i = \int \overline{\eta}_i d\overline{\eta}_i = i, \qquad [d\overline{\eta}_i, d\eta_j] = 0 \qquad \{d\eta_j, \eta_i\} = \{d\eta_j, a_i^{\pm}\} = 0$$

Completeness conditions are:

$$1 = i \int |\eta\rangle \langle \eta| d^2\eta = -i \int d^2\overline{\eta} |\overline{\eta}\rangle \langle \overline{\eta}| \quad d^2\eta = d\eta_2 d\eta_1 \quad d^2\overline{\eta} = d\overline{\eta}_1 d\overline{\eta}_2$$

Important formula for us in the coherent state formalism is the trace formula enabling us to calculate trace of some U matrix, which consists of γ matrices, as an integral over the Grassmann variables:

$$Tr\left(U\right) = i \int d^2\eta \; \langle -\eta| U |\eta\rangle \quad (19)$$

2.2 Worldline Formalism for the Spinor Loop

Let us denote the operator in (16) by Σ:

$$\Sigma = -\left(D^2 - \frac{1}{4} i g F_{\mu\nu} \left[\gamma_\mu, \gamma_\nu\right]\right); \quad \Gamma(A) = \frac{1}{2} Tr \log(-\Sigma + m^2)$$

Then using (5) we can write the effective action (16) as an exponent and apply the formula (19)

$$Tr\ e^{-T\Sigma} = i\int d^4x \int d^2\eta\ \langle x,-\eta|\, e^{-T\Sigma}\,|x,\eta\rangle \tag{20}$$

For spinor loop effective action we additionaly suppose that Grassmann variables is also defined on the worldloop, i.e. are functions of τ. So, on dividing T into N equal parts the matrix element (20) becomes product of following ones:

$$i^N \int d^4x^i\ d^2\eta^i\ \langle x^i,\eta^i|\, e^{-\frac{T}{N}\Sigma}\,|x^{i+1},\eta^{i+1}\rangle$$

Boundary conditions for pairs of variables (x,η) are $(x^{N+1},\eta^{N+1})=(x^1,-\eta^1)$. These are the antiperiodic boundary condition for η states associated with the definition of trace via Grassman integral (19). For example, the matrix element of the product of two γ matices is expressed by means of following Grassmann integrals

$$\langle\eta^i|\,\gamma_\mu\gamma_\nu\,|\eta^{i+1}\rangle = -i\int d^2\overline{\eta}^{i,i+1}\,\langle\eta^i|\,|\overline{\eta}^{i,i+1}\rangle\,\langle\overline{\eta}^{i,i+1}|\,|\eta^{i+1}\rangle\, 2\ {}^i\psi_\mu\psi_\nu^{i+1}, \mu\neq\nu,$$

where we have introduced new Grassmann variables ψ_m as:

$$\psi_{1,2}^{i+1} \equiv \frac{1}{\sqrt{2}}\left(\eta_{1,2}^{i+1}+\overline{\eta}_{1,2}^{i,i+1}\right) \qquad {}^i\psi_{1,2} \equiv \frac{i}{\sqrt{2}}\left(\eta_{1,2}^{i}+\overline{\eta}_{1,2}^{i,i+1}\right)$$

$$\psi_{3,4}^{i+1} \equiv \frac{1}{\sqrt{2}}\left(\eta_{1,2}^{i+1}-\overline{\eta}_{1,2}^{i,i+1}\right) \qquad {}^i\psi_{3,4} \equiv \frac{i}{\sqrt{2}}\left(\eta_{1,2}^{i}-\overline{\eta}_{1,2}^{i,i+1}\right).$$

Thereby the matrix elements which constitute the fermionic effective action is

$$\langle x^i,\eta^i|\, e^{-\frac{T}{N}\Sigma(p,A,\gamma_\mu\gamma_\nu)}\,|x^{i+1},\eta^{i+1}\rangle = -\frac{i}{(2\pi)^4}\int d^4p^{i,i+1}\ d^2\overline{\eta}^{i,i+1}$$

$$\times e^{i(x^i-x^{i+1})p^{i,i+1}+(\eta^i-\eta^{i+1})_r\overline{\eta}_r^{i,i+1}}\left\{1-\frac{T}{N}\sum\left(p^{i,i+1},\ A^{i,i+1}, 2\ {}^i\psi_\mu\psi_\nu^{i+1}+O\left(\frac{T^2}{N^2}\right)\right)\right\}$$

and trace (20) will consist of the following matrix elements

$$Tr\ e^{-T\Sigma} = \frac{1}{(2\pi)^{4N}}\int\prod_{I=1}^{N} d^4x^i\ d^4p^{i,i+1}\ d^2\eta^{i,i+1}\ d^2\overline{\eta}^{i,i+1}\left(1-\frac{T}{N}\Sigma+O\left(\frac{T^2}{N^2}\right)\right)$$

$$\times\exp\left\{\sum_{i=1}^{N}\left[i\left(x^i-x^{i+1}\right)p^{i,i+1}+\frac{1}{2}\left(\eta^i-\eta^{i+1}\right)_r\overline{\eta}_r^{i,i+1}-\frac{1}{2}\eta_r^i\left(\overline{\eta}_r^{i-1,i}-\overline{\eta}_r^{i,i+1}\right)\right]\right\}.$$

On going to the limit $N\to\infty$ we have additional path integrals over Grassmann variables, which obey antiperiodic boundary conditions:

$$Tr\ e^{-T\Sigma}=tr\int Dp\int Dx\int_A D\eta D\overline{\eta}\ \exp\left\{\int_0^T d\tau\left[ix\,\dot{p}+\frac{1}{2}\,\dot{\eta}_r\,\overline{\eta}_r-\frac{1}{2}\eta\,\dot{\overline{\eta}}_r-\Sigma\left(p,\ A,2\psi_\mu\psi_\nu\right)\right]\right\}.$$

Here tr means trace over the color indices. It is useful to write Grassmann path integral in terms of the ψ variables

$$\psi_{1,2}(\tau) = \frac{1}{\sqrt{2}}\left(\eta_{1,2}(\tau) + \overline{\eta}_{1,2}(\tau)\right), \qquad \psi_{3,4}(\tau) = \frac{i}{\sqrt{2}}\left(\eta_{1,2}(\tau) - \overline{\eta}_{1,2}(\tau)\right)$$

for which we can write

$$\frac{1}{2}\dot{\eta}_r\,\overline{\eta}_r - \frac{1}{2}\eta\,\dot{\overline{\eta}}_r = -\frac{1}{2}\psi^\mu\,\dot{\psi}_\mu$$

and boundary condition becomes:

$$(x(T),\ \psi(T)) = (x(0),\ -\psi(0))$$

Finally the spinor loop effective action is given by the following path integral expression:

$$\Gamma_{spin}[A] = -tr\frac{1}{2}\int_0^\infty \frac{dT}{T}e^{-m^2T}\int_P Dx\int_A D\psi\, e^{-\int_0^T d\tau L_{spin}}, \tag{21}$$

where L_{spin} denotes

$$L_{spin} = \frac{1}{4}\dot{x}^2 + \frac{1}{2}\psi^\mu\,\dot{\psi}_\mu + igA_\mu\,\dot{x}^\mu - ig\psi^\mu F_{\mu\nu}\psi^\nu.$$

We should make the formula (21) applicable for the calculation of the one loop N-point amplitudes. To this end let us expand Γ_{spin} and ignore the term $[A_\mu, A_\nu]$ in $F_{\mu\nu}$. Then N-th term in this expansion will be equal to:

$$\Gamma[A] = -\frac{1}{2}\frac{(ig)^N}{N!}\int_0^\infty \frac{dT}{T}e^{-m^2T}\int_P Dx\int_A D\psi\,\exp\left[-\int_0^T d\tau\left\{\frac{1}{2\mathcal{E}}\dot{x}^2 + \frac{1}{2}\psi^\mu\,\dot{\psi}_\mu\right\}\right]$$
$$\times\, tr\prod_{i=1}^N\int_0^T dt_i\left\{A_\mu[x(t_i)]\cdot\dot{x}^\mu(t_i) - \mathcal{E}\psi^\mu(t_i)\,\partial_\mu A_\nu[x(t_i)]\,\psi^\nu(t_i)\right\}$$

For the plane wave choice of A^μ, it becomes

$$= -\frac{1}{2}\frac{(ig)^N}{N!}\int_0^\infty \frac{dT}{T}e^{-m^2T}\int_P Dx\int_A D\psi\,\exp\left[-\int_0^T d\tau\left\{\frac{1}{2\mathcal{E}}\dot{x}^2 + \frac{1}{2}\psi^\mu\,\dot{\psi}_\mu\right\}\right]$$
$$\times\, tr\prod_{i=1}^N T^{a_i}\int_0^T dt_i\left\{\epsilon_i\cdot\partial_i x(t_i) + i\mathcal{E}\epsilon_i\cdot\psi(t_i)\,k_i\cdot\psi(t_i)\right\}e^{ik_i\cdot x(t_i)}. \tag{22}$$

Again we have non-exponent bracket, which does not permit us to carry out the next transformation. We can however write it as an exponent in the following way

$$V \equiv igT^a \left[\epsilon \cdot \dot{x} + i\mathcal{E}\epsilon \cdot \psi \; k \cdot \psi\right] e^{ik\cdot x}$$
$$= igT^a \int d\theta d\overline{\theta} \; \exp\left[\overline{\theta}_i \theta_i \epsilon \cdot \dot{x} + \theta\sqrt{\mathcal{E}}\epsilon \cdot \psi + i\overline{\theta}\sqrt{\mathcal{E}} k \cdot \psi + ik \cdot x\right]$$

and the bracket in this exponent can be divided into two "currents"

$$J^\mu(\tau) = \sum_1^N \delta(\tau - t_i)\left(\overline{\theta}_i \theta_i \epsilon_i^\mu \partial_{t_i} + ik_i^\mu\right),$$

$$\eta^\mu\left(\tau, \theta, \overline{\theta}\right) = \sum_1^N \delta(\tau - t_i)\left(\theta_i \epsilon_i^\mu + i\overline{\theta}_i k_i^\mu\right).$$

The result of carrying out the x and ψ integrals in the $\mathcal{E} = 2$ gauge gives us main worldline formula for spinor loop contribution to N-point amplitude:

$$\begin{aligned}
\Gamma_N^0(k_1, \ldots k_N) = &-4\frac{(ig)^N}{2(4\pi)^2} Tr\left(T^{a_N} \ldots T^{a_1}\right) \int_0^\infty \frac{dT}{T^{3-N}} \left(\prod_{i=1}^N \int_0^{u_{i+1}} du_i\right) \\
&\times \left[\sum_{i<j=1}^N k_i \cdot k_j G_B^{ji}\right] \left\{ \left(\prod_{i=1}^N \int d\theta_i d\overline{\theta}_i\right)\right. \\
&\times \exp\left(\sum_{i<j=1}^N \left[-i\left(\overline{\theta}_j \theta_j k_i \cdot \epsilon_i - \overline{\theta}_i \theta_i k_j \cdot \epsilon_i\right) \dot{G}_B^{ji} + \overline{\theta}_i \theta_i \overline{\theta}_j \theta_j \epsilon_i \cdot \epsilon_j \ddot{G}_B^{ji}\right]\right) \\
&\times \exp\left.\left(\sum_{i<j=1}^N \left[-\overline{\theta}_i \overline{\theta}_j k_i \cdot k_j + i\overline{\theta}_i \theta_j k_i \cdot \epsilon_j + i\theta_i \overline{\theta}_j \epsilon_i \cdot k_j + \theta_i \theta_j \epsilon_i \cdot \epsilon_j\right] G_F^{ji}\right)\right\}.
\end{aligned} \tag{23}$$

Trace of the unit matrix is equal to

$$4 = \int_A D\psi \; \exp\left[-\int_0^T d\tau \left\{\frac{1}{2}\psi^\mu \; \dot{\psi}_\mu\right\}\right] = tr_\psi 1 = \sum_{\alpha=1}^4 \langle\alpha \,|\alpha\rangle.$$

In obtaining (23), we have introduced fermionic Green's function G_F as:

$$\frac{1}{2}\partial_t G_F(t, t') = \delta(t - t')$$

obeying antiperiodic boundary condition

$$\psi(t \to T) = -\psi(t \to 0)$$

This equation has a solution:

$$G_F(t,t') = sign(t-t') = sign(u-u')$$

We neglected the term $-\frac{1}{2}g^2\mathcal{E}\psi^\mu[A_\mu, A_\nu]\psi^\nu$ when carrying out the transformations, and the formula (23) does not contain contribution of this term. This term for plane wave A^μ will have the form:

$$O_{i,j} = -g^2\mathcal{E}\left(T^{a_j}T^{a_i}\right)\epsilon_j\cdot\psi\;\epsilon_i\cdot\psi\;e^{i(k_i+k_j)\cdot x}.$$

and after writing polarization vectors in the exponent it becomes integrals over the Grassmann variables:

$$O_{i,j} = -g^2\mathcal{E}\left(T^{a_j}T^{a_i}\right)\int d\theta_i d\bar{\theta}_i d\theta_j d\bar{\theta}_j\;\left(-\bar{\theta}_i\bar{\theta}_j\right)$$

$$\times\exp\left[\theta_i\sqrt{\mathcal{E}}\;\epsilon_i\cdot\psi(t_i) + \theta_j\sqrt{\mathcal{E}}\;\epsilon_j\cdot\psi(t_j) + i(k_i+k_j)\cdot x\right]|_{t_i=t_j}\,.$$

Setting $N=2$ in the (23) we obtain spinor loop contribution to two point amplitude - gluon polarization tensor:

$$\Gamma_2(k_1,k_2) = -2\frac{\left(ig\mu^{\epsilon/2}\right)^2}{(4\pi)^{2-\epsilon/2}}Tr\left(T^aT^b\right)\int\limits_0^\infty\frac{dT}{T^{1-\epsilon/2}}\int\limits_0^1 du\;e^{k_1\cdot k_2 G_B(1-u)}$$

$$\Big\{k_2\cdot\epsilon_1 k_1\cdot\epsilon_2\left[\dot{G}_B(1-u)\right]^2 + \epsilon_1\cdot\epsilon_2\,\ddot{G}_B(1-u)$$

$$+\left(\epsilon_1\cdot\epsilon_2 k_1\cdot k_2 - k_2\cdot\epsilon_1 k_1\cdot\epsilon_2\right)\left[G_F(1-u)\right]^2\Big\}.$$

It can be checked that this expression coincides with the one obtained by direct calculation in proper time parametrization:

$$\Pi(k_1,k_2) = 2\frac{\left(g\mu^{\epsilon/2}\right)^2}{(4\pi)^{2-\epsilon/2}}Tr\left(T^aT^b\right)\int\limits_0^\infty\frac{dT}{T^{1-\epsilon/2}}\left\{\int\limits_0^1 da\;e^{-Tk_1^2\left(a-a^2\right)}\right.$$

$$\left.\times\left[-\frac{2}{T}\epsilon_1\cdot\epsilon_2 + (1-2a)^2 k_2\cdot\epsilon_1 k_1\cdot\epsilon_2 + \left(\epsilon_1\cdot\epsilon_2 k_1\cdot k_2 - k_2\cdot\epsilon_1 k_1\cdot\epsilon_2\right)\right] + \frac{2}{T}\epsilon_1\cdot\epsilon_2\right\}.$$

LECTURE 2

AdS/CFT Correspondence of Correlators

3 Worldline Formalism for the Free Scalar Field Theory

As we have noted before, the $\mathcal{N}=4$ Super Yang-Mills theory at zero coupling is dual to string theory on highly curved *AdS* space. Since first theory is obtained as the $\alpha' \to 0$ limit of open string theory, the planar open string worldsheets will reduce to planar worldlines in this limit. More precisely, the worldsheet moduli space integral reduces to a worldline moduli space integral known as a Schwinger parameter integral for the Feynman diagrams. On the other hand, AdS supergravity can be reduced from closed string theory and correlators in this theory also can be written in parameter integrals. It would be reasonable to realize closed/open string duality by means of correlation functions in AdS supergravity and free field theory [6]. Worldline formalism turned out to be useful for the realization of this idea also. Simplest case of free field theory is free scalar field theory. Let us start by showing the equivalence of amplitudes in the open and closed string theories.

3.1 General Formulas

Let us consider a Euclidean free scalar field theory in d dimension. One loop N-point function of the gauge invariant operator $Tr\Phi^2$ is equal to

$$\Gamma\left(x_1, x_2, ...x_n\right) = \frac{1}{N^2}\left\langle Tr\Phi^2\left(x_1\right) Tr\Phi^2\left(x_2\right) \cdots \cdot Tr\Phi^2\left(x_n\right)\right\rangle$$

In string theory analogous amplitude is given by following expression:

$$\Gamma\left(k_1, k_2, ...k_n\right) = \int [dM] \left\langle e^{ik_1\cdot X(\tau_1)} e^{ik_2\cdot X(\tau_2)} \ldots e^{ik_n\cdot X(\tau_n)}\right\rangle$$

$$\equiv \int_0^\infty \frac{d\tau}{\tau} \int_0^\tau \prod_{i=1}^n d\tau_i \left\langle e^{ik_1\cdot X(\tau_1)} e^{ik_2\cdot X(\tau_2)} \ldots e^{ik_n\cdot X(\tau_n)}\right\rangle, \quad (24)$$

where $[dM]$ is string worldsheet element and parameter τ is Schwinger parameter denoted as T in the previous lecture. The correlator of worldline vertex operators is expressed by means of Green's function on the circle [5]

$$\left\langle e^{ik_1\cdot X(\tau_1)} e^{ik_2\cdot X(\tau_2)} \ldots e^{ik_n\cdot X(\tau_n)}\right\rangle = \delta^{(d)}\left(\sum_i k_i\right) \frac{1}{\tau^{d/2}} \exp\left(-\frac{1}{2}\sum_{i\neq j}^n k_i \cdot k_j G\left(\tau_i, \tau_j\right)\right) \quad (25)$$

$$G(\tau_i,\tau_j) = -\frac{\tau_{ij}(\tau-\tau_{ij})}{\tau}; \qquad \tau_{ij} = |\tau_i - \tau_j|$$

N-point function in scalar field theory has expression:

$$\Gamma(k_1,k_2,....k_n) = \delta^{(d)}\left(\sum_i k_i\right)\int_0^\infty \frac{d\tau}{\tau^{d/2+1}}\int_0^\tau \prod_{i=1}^n d\tau_i \exp\left(-\frac{1}{2}\sum_{i\neq j}^n k_i\cdot k_j G(\tau_i,\tau_j)\right). \tag{26}$$

3.2 The Two Point Function

The two point function is obtained from the formula (26) by setting $n=2$, taking into account energy-momentum conservation in the exponent and making a change of variable $\tau_{12}=\tau\alpha$

$$\Gamma(k_1,k_2) = \delta^{(d)}(k_1+k_2)\int_0^\infty \frac{d\tau}{\tau^{d/2+1}}\tau^2\int_0^1 d\alpha e^{-\tau\alpha(1-\alpha)\left(\beta k_1^2+(1-\beta)k_2^2\right)} \tag{27}$$

Here the parameter β has been introduced by hand ($0\le\beta\le1$) one can check that (27) does not depend on this parameter. Two point function in position space can be obtained by Fourier transformation

$$\Gamma(x_1,x_2) = \int\prod_{i=1}^2 d^d k_i e^{-ik_i\cdot x_i}\Gamma(k_1,k_2)$$

$$\tilde{\Gamma}(k) = \int_0^\infty \frac{d\tau}{\tau^{d/2+1}}\tau^2\int_0^1 d\alpha e^{-\tau\alpha(1-\alpha)k^2}$$

Using the formula

$$\langle x\left|\exp[t\Box]\right|y\rangle = \frac{1}{(4\pi t)^{d/2}}e^{-\frac{(x-y)^2}{4t}}, \tag{28}$$

$\Gamma(x_1,x_2)$ can be written in terms of heat kernel as:

$$\begin{aligned}\Gamma(x_1,x_2) &= \int_0^\infty \frac{d\tau}{\tau^{d/2+1}}\tau^2\int d^d z\int_0^1 d\alpha\left\langle x_1\left|e^{\tau\alpha(1-\alpha)\beta\Box}\right|z\right\rangle\left\langle z\left|e^{\tau\alpha(1-\alpha)(1-\beta)\Box}\right|x_2\right\rangle\\ &= \int_0^\infty \frac{d\tau}{\tau^{d/2+1}}\tau^2\int_0^1 d\alpha\left\langle x_1\left|e^{\tau\alpha(1-\alpha)\beta\Box}\right|x_2\right\rangle \end{aligned}\tag{29}$$

The two point amplitude in *AdS* space is expressed by means of bulk contribution to boundary propagator in the following way

$$\Gamma\left(x_1,x_2\right)=\int d^dz\int\limits_0^\infty \frac{dz_0}{z_0^{d+1}}\prod_{i=1}^2 K\left(x_i,z;z_0\right).$$

In terms of the heat kernel, this can be written as:

$$\Gamma\left(x_1,x_2\right)=\int\limits_0^\infty \frac{dt}{t^{d/2+1}}t^2\int\limits_{-\infty}^\infty d^dz\int\limits_0^\infty d\rho_1 d\rho_2\left(\rho_1\rho_2\right)^{d/2-3}$$
$$\times\, e^{-\rho_1-\rho_2}\left\langle x_1\left|e^{\frac{t}{4\rho_1}\Box}\right|z\right\rangle\left\langle x_2\left|e^{\frac{t}{4\rho_2}\Box}\right|z\right\rangle, \qquad (30)$$

where $t=z_0^2$. By changing the variables as

$$\rho_1=\rho\left(1-\beta\right),\quad \rho_2=\rho\beta$$

and

$$t=4\tau\rho\alpha\left(1-\alpha\right)\left(1-\beta\right)\beta$$

we get

$$\Gamma\left(x_1,x_2\right)=\int\limits_0^1\frac{d\beta}{\beta\left(1-\beta\right)}\int\limits_0^\infty\frac{d\tau}{\tau^{d/2+1}}\tau^2\int d^dz$$

$$\times\int\limits_0^1\frac{d\alpha}{\left[\alpha\left(1-\alpha\right)\right]^{d/2-2}}\left\langle x_1\left|e^{\tau\alpha(1-\alpha)\beta\Box}\right|z\right\rangle\left\langle z\left|e^{\tau\alpha(1-\alpha)(1-\beta)\Box}\right|x_2\right\rangle$$

$$=\int\limits_0^1\frac{d\beta}{\beta\left(1-\beta\right)}\int\limits_0^\infty\frac{d\tau}{\tau^{d/2+1}}\tau^2\int\limits_0^1\frac{d\alpha}{\left[\alpha\left(1-\alpha\right)\right]^{d/2-2}}\left\langle x_1\left|e^{\tau\alpha(1-\alpha)\beta\Box}\right|x_2\right\rangle \qquad (31)$$

3.3 The Three Point Function

To get three point function we should set $N=3$ in the formula (26):

$$\Gamma\left(k_1,k_2,k_3\right)=\delta^{(d)}\left(k_1+k_2+k_3\right)\int\limits_0^\infty\frac{d\tau}{\tau^{d/2+1}}\int\limits_0^\tau\prod_{i=1}^3 d\tau_i\exp\left(-\frac{1}{2}\sum_{i\neq j}^3 k_i\cdot k_j G\left(\tau_i,\tau_j\right)\right)$$

Again we use energy-momentum conservation in the exponent to express $k_i\cdot k_j$ factors as sums of k_l^2 as we made in two point case. For example,

$$2k_1\cdot k_2=k_3^2-k_1^2-k_2^2$$

Doing change of variables

$$\tau_{12}=\tau\alpha_3,\ \tau_{23}=\tau\alpha_1,\ \tau_{31}=\tau\alpha_2$$

we get the following expression for the $\Gamma(k_1, k_2, k_3)$

$$\Gamma(k_1, k_2, k_3) = \delta^{(d)}(k_1 + k_2 + k_3) \int_0^\infty \frac{d\tau}{\tau^{d/2+1}} \tau^3 \int_0^\tau \prod_{i=1}^3 d\alpha_i \delta\left(\sum_i \alpha_i - 1\right)$$

$$\times \exp\left[-\tau\left(k_1^2 \alpha_2 \alpha_3 + k_2^2 \alpha_3 \alpha_1 + k_3^2 \alpha_1 \alpha_2\right)\right],$$

where we have inserted $\delta\left(\sum_i \alpha_i - 1\right)$ by hand, which follows from the equality $\tau_{12} + \tau_{23} + \tau_{31} = \tau$. We can convert exponent into the heat kernel form by use of formula (5)

$$\exp\left[-\tau \sum_i k_i^2 \alpha_j \alpha_k\right] = \prod_{i=1}^3 \langle k_i | \exp[\tau \alpha_j \alpha_k \Box] | k_i \rangle,$$

After carrying out momentum integrals in position space, three-point function can be expressed in terms of heat kernel as well:

$$\Gamma(x_1, x_2, x_3) = \int \prod_{i=1}^3 d^d k_i e^{-ik_i \cdot x_i} \Gamma(k_1, k_2, k_3)$$

$$= \int_0^\infty \frac{d\tau}{\tau^{d/2+1}} \tau^3 \int_0^\tau \prod_{i=1}^3 d\alpha_i \delta\left(\sum_i \alpha_i - 1\right) \frac{1}{\tau^{3d/2}\left(\prod \alpha_i\right)^d} \int d^d z \exp\left\{-\frac{1}{4}\sum_{i=1}^3 \frac{(x_i - z)^2}{\tau \alpha_j \alpha_k}\right\}$$

$$= \int_0^\infty \frac{d\tau}{\tau^{d/2+1}} \tau^3 \int_0^\tau \prod_{i=1}^3 d\alpha_i \delta\left(\sum_i \alpha_i - 1\right) \int d^d z \prod_{i=1}^3 \langle x_i | \exp[\tau \alpha_j \alpha_k \Box] | z \rangle \quad (32)$$

3.4 Bulk to Boundary Propagator for Scalar Field in *AdS* Space

In Euclidean AdS_{d+1} space with the boundary at $z_0 = 0$ in the Poincare coordinates

$$ds^2 = \frac{d\vec{z}^2 + dz_0^2}{z_0^2}$$

the wave equation has following form

$$\left[-z_0^2 \partial_{z_0}^2 + (d-1) z_0 \partial_{z_0} - z_0^2 \Box - 2(d-2)\right] K = 0, \quad (33)$$

where $\Box$ is the d-dimensional Laplacian in the $\vec{z}$ direction. In terms of coordinate $t = z_0^2$ and defining $K = t\widetilde{K}$, equation (33) takes the form

$$\left[2(d-6)\frac{\partial \widetilde{K}}{\partial t} - \Box \widetilde{K} - 4t \frac{\partial^2 \widetilde{K}}{\partial t^2}\right] = 0$$

The solution of this equation has the form:

$$\widetilde{K}(t) = \int_0^\infty d\rho \rho^{d/2-3} e^{-\rho} e^{\frac{t}{4\rho}\Box} \tag{34}$$

Bulk to boundary propagator with $\triangle = d-2$, finally gets the explicit form:

$$K\left(x, z; z_0 = t^{1/2}\right) = \left(\frac{t^{1/2}}{t + (x-z)^2}\right)^{d-2} = \langle x \left| K(t) \right| z \rangle$$

3.5 The Three Point Amplitude in *AdS* Space

The three point amplitude in *AdS* space is expressed by means of bulk to boundary propagator in the following way

$$\Gamma(x_1, x_2, x_3) = \int d^d z \int_0^\infty \frac{dz_0}{z_0^{d+1}} \prod_{i=1}^3 K(x_i, z; z_0). \tag{35}$$

Having expressed in terms of $\widetilde{K}(t)$ function it can be written as integrals of ρ_i

$$\Gamma(x_1, x_2, x_3) = \int_0^\infty \frac{dt}{t^{d/2+1}} t^3 \int_0^\infty \prod_{i=1}^3 d\rho_i \rho_i^{d/2-3} e^{-\rho_i} \int d^d z \prod_{i=1}^3 \left\langle x_i \left| e^{\frac{t}{4\rho_i}\Box} \right| z \right\rangle \tag{36}$$

By changing the variables as

$$\rho_i = \rho \alpha_i, \qquad \int_0^\infty d\rho \delta\left(\rho - \sum_i \rho_i\right) = 1, \quad t = 4\tau\rho \prod_{i=1}^3 \alpha_i \tag{37}$$

the amplitude (36) will coincide with the three-point function (32) we obtained before. This demonstrates equality of three point function of free scalar field theory to the one in *AdS* supergravity and prove the idea about duality of field theory and supergravity in *AdS* space.

4 AdS/CFT Correspondence of *R*-current Two Point Correlator

As we noted *AdS*/*CFT* duality is a consequence of the closed/open string duality. The SYM theory on the boundary and supergravity in the bulk are effective theories of open and closed strings, respectively. Using world-line formalism we have shown how to convert correlators of one theory into another one for the free scalar theory. The correspondence of correlators of R-symmetry currents in $\mathcal{N} = 4$ super Yang-Mills theory and supergravity correlators was established in [7]. in that work, where two and three point correlators were related with each other and the AdS/CFT correspondence was confirmed for this as well. Having gained experience in the free scalar field case we can now extend the ideas of [7] in the worldline formalism framework. Namely, we will use the procedure of introducing bulk to boundary propagators via parameter integrals of heat kernel in the two point R-current correlation function in super Yang-Mills theory and thereby rewrite R-current correlator in terms of bulk to boundary propagator and match with supergravity correlator. Supersymmetry in these currents enables us to use them in the free field limit, i.e. consider only contributions of scalar and spinor loops and neglect gluon loop contribution.

4.1 The Set-up in *AdS* Space

Let us start with the *AdS* supergravity fields which are the corresponding partners of SYM R-currents. These *AdS* fields are $SU(4)$ gauge fields with the following action [7]:

$$S[A] = \frac{1}{2g^2}\int \delta_{ab} dA^a \wedge *dA^b + f_{abc} dA^a \wedge *\{A^b \wedge A^c\} + \frac{ik}{32\pi^2}\int d_{abc} A^a \wedge dA^b \wedge dA^c \,. \tag{38}$$

where integrals are over 5-dimensional *AdS*-space and the last term is the Chern-Simons term.

Solutions of the classical equations of motion would be determined in terms of their boundary values, $a_i^a(\boldsymbol{x})$ as follows [7]:

$$A^a(z_0,\boldsymbol{z}) = dz_i \int d^d x a_i^a(\boldsymbol{x}) \frac{z_0^{d-2}}{R(x;z_0,z)^{d-1}} - dz_0 \int d^d x a_i^a(\boldsymbol{x}) \frac{1}{2(d-2)} \frac{\partial}{\partial x^i} \frac{z_0^{d-3}}{R(x;z_0,z)^{d-2}} \tag{39}$$

where $R(x;z_0,z) = z_0^2 + |\boldsymbol{z} - \boldsymbol{x}|^2$. Plugging this solution in the action will give us the effective action in tree level as a functional of boundary values [7]:

$$S[a] = \frac{1}{2g^2}\int d^d x_1 d^d x_2 a_i^a(x_1) a_j^b(x_2) \delta_{ab} \times \left\{ (\partial_1^i \partial_1^j - \delta^{ij}\partial_1^2) I^{d/2-1}_{d-2,d-2} - \frac{1}{8(d-2)^2} (\partial_1^i \partial_1^j - \delta^{ij}\partial_1^2)\partial_1^2 I^{d/2-2}_{d-3,d-3} \right\} + \frac{1}{2} g^2 \int d^d x_1 d^d x_2 d^d x_3 a_i^a(x_1) a_j^b(x_2) a_k^c(x_3) f_{abc}$$

$$\times \left\{ 2\delta^{ij}\partial_{1k} I^{d-2}_{d-2,d-2,d-2} + \frac{1}{2(d-2)^2}\partial_{2j}(\partial^i_1\partial^k_1 - \delta^{ik}\partial^2_1) I^{d-3}_{d-3,d-3,d-2} \right\}$$
$$+ \frac{ik}{32\pi^2}\int d^4x_1 d^4x_2 d^4x_3 a^a_i(x_1) a^b_j(x_2) a^c_k(x_3) d_{abc}\epsilon^{jklm}\frac{1}{d-2}$$
$$\times \left\{ -\frac{1}{2}\partial_{1i}\partial_{2m}\partial_{3l} I^{3d/2-4}_{d-3,d-2,d-2} + \partial_{3l}(\partial_{1i}\partial_{1m} - \delta^{im}\partial^2_1) I^{3d/2-4}_{d-3,d-2,d-2} \right\}, \quad (40)$$

where,

$$I^f_{mn}(x_1,x_2) = \int dz_0 d^d z \frac{z_0^{2f+1}}{R(x_1;z_0,z)^{m+1} R(x_2;z_0,z)^{n+1}} \quad (41)$$
$$I^f_{mnp}(x_1,x_2,x_3) = \int dz_0 d^d z \frac{z_0^{2f+1}}{R(x_1;z_0,z)^{m+1} R(x_2;z_0,z)^{n+1} R(x_3;z_0,z)^{p+1}} . \quad (42)$$

In the tree level the above solution is in zeroth order in the Yang-Mills coupling and there is no need for ghosts. Two-point function of the R-currents is found by taking functional derivatives from this generating function with respect to boundary fields $a^a_i(\boldsymbol{x})$:

$$\left\langle J^i_a(x_1) J^j_b(x_2) \right\rangle = -\frac{1}{g^2}\delta_{ab}(\partial^i_1\partial^j_1 - \delta^{ij}\partial^2_1) I^1_{22} \, . \quad (43)$$

This is the correlation function derived from the supergravity theory in the bulk.

4.2 Two-point Correlator for the Boundary Theory

Beside the scalar and spinor loops, in super Yang-Mills theory there is a vector field loop, which doesn't contribute to this correlator in the free field limit.

Worldline expression of scalar loop contribution to this function was obtained from (14) with $N = 2$ in previous lecture, which after a little rearrangement gets the following well known form (15):

$$\Pi^{ab}(k_1,k_2) = -Tr\left(T^a T^b\right)\frac{\left(g\mu^{\epsilon/2}\right)^2}{(4\pi)^{2-\epsilon/2}}\int_0^\infty \frac{d\tau}{\tau^{d/2-1}}\int_0^1 d\alpha\, e^{\tau k_1 k_2 \alpha(1-\alpha)}\int_{-\infty}^{\infty}\frac{d^d z}{(2\pi)^d} e^{i(k_1+k_2)z}$$
$$\times \left[\frac{2}{\tau}\left(\delta(\alpha)-1\right)\epsilon_1\cdot\epsilon_2 + (1-2\alpha)^2(\epsilon_1\cdot k_2)(\epsilon_2\cdot k_1)\right] .$$

Note that the polarization operator-two point function (34) contains contribution of tadpole diagram of vector-scalar interaction in $\delta(\alpha)$-function term. Comparing this correlator with one in free scalar field theory (27) we find that the expression under integration differs from the two-point correlation function of scalar field only by additional square bracket factor in it. Now let us go to position space by means of the inverse Fourier transformation:

$$\Pi^{ab}(x_1,x_2) = -Tr\left(T^aT^b\right) C \int\limits_0^\infty \frac{d\tau}{\tau^{d/2-1}} \int\limits_{-\infty}^\infty d^dk_1 \int\limits_{-\infty}^\infty d^dk_2 \int\limits_{-\infty}^\infty \frac{d^dz}{(2\pi)^d} e^{i(k_1+k_2)z}$$

$$\times \int\limits_0^1 d\alpha \int\limits_0^1 d\beta \; e^{-\tau k_1^2\beta\alpha(1-\alpha)-\tau k_2^2(1-\beta)\alpha(1-\alpha)} e^{-ik_1x_1-ik_2x_2}$$

$$\times \left[\frac{2}{\tau}\left(\delta(\alpha)-1\right)\epsilon_1\cdot\epsilon_2 + (1-2\alpha)^2\left(\epsilon_1\cdot\left(i\frac{\partial}{\partial x_2}\right)\right)\left(\epsilon_2\cdot\left(i\frac{\partial}{\partial x_1}\right)\right)\right] \quad (44)$$

Here we have inserted an integral over the parameter β, where $0 \le \beta \le 1$. Taking Gaussian integrals first over the momenta and then the position derivatives, we obtain from (44) the following expression for polarization operator[1]:

$$\Pi^{ab}(x_1,x_2) = -CTr\left(T^aT^b\right) \int\limits_0^\infty \frac{d\tau}{\tau^{d/2-1}} \int\limits_0^1 d\alpha \int\limits_0^1 d\beta \int\limits_{-\infty}^\infty d^dz \left(\frac{\pi}{\tau\beta\alpha(1-\alpha)}\right)^{d/2}$$

$$\times e^{-\frac{(z-x_1)^2}{4\tau\beta\alpha(1-\alpha)}} \left(\frac{\pi}{\tau(1-\beta)\alpha(1-\alpha)}\right)^{d/2} e^{-\frac{(z-x_2)^2}{4\tau(1-\beta)\alpha(1-\alpha)}} \left[\frac{2}{\tau}\left(\delta(\alpha)-1\right)\epsilon_1\cdot\epsilon_2\right.$$

$$\left. -(1-2\alpha)^2 \frac{1}{2\tau(1-\beta)\alpha(1-\alpha)}\epsilon_1\cdot(z-x_2)\frac{1}{2\tau\beta\alpha(1-\alpha)}\epsilon_2\cdot(z-x_1)\right]. \quad (45)$$

Writing the exponents in heat kernel terms by means of (28) we rewrite (45) as follows:

$$\Pi^{ab}(x_1,x_2) = -C'Tr\left(T^aT^b\right) \int\limits_0^\infty \frac{d\tau}{\tau^{d/2-1}} \int\limits_0^1 d\alpha \int\limits_0^1 d\beta \int\limits_{-\infty}^\infty d^dz \left\langle x_1 \left| e^{\tau\beta\alpha(1-\alpha)\Box} \right| z\right\rangle \quad (46)$$

$$\times \left\langle x_2 \left| e^{\tau(1-\beta)\alpha(1-\alpha)\Box} \right| z\right\rangle \left[\frac{2}{\tau}\left(\delta(\alpha)-1\right)\epsilon_1\cdot\epsilon_2 - (1-2\alpha)^2 \frac{\epsilon_1\cdot(z-x_2)}{2\tau\alpha(1-\beta)(1-\alpha)} \frac{\epsilon_2\cdot(z-x_1)}{2\tau\beta\alpha(1-\alpha)}\right].$$

We can insert the integral representation of $\Gamma(s)$ function,

$$1 = \frac{1}{\Gamma(d/2)} \int\limits_0^\infty d\rho\rho^{d/2-1}e^{-\rho} = \frac{1}{\Gamma(d/2+1)} \int\limits_0^\infty d\rho\rho^{d/2}e^{-\rho}, \quad (47)$$

into (46) and pass to new variables used in previous subsection:

$$\rho_1 = \rho(1-\beta), \; \rho_2 = \beta\rho \quad \text{and} \quad t = 4\tau\rho\beta(1-\beta)\alpha(1-\alpha)$$

[1] We have included constants into the new one.

with $\int\limits_0^\infty \rho d\rho \int\limits_0^1 d\beta = \int\limits_0^\infty d\rho_1 \int\limits_0^\infty d\rho_2$, then the polarization operator (46) is rewritten in more symmetric a form:

$$\Pi^{ab}(x_1,x_2) = \int\limits_0^\infty \frac{dt}{t^{\frac{d}{2}+1}} \int\limits_{-\infty}^\infty d^d z \int\limits_0^\infty d\rho_1 \rho_1^{\frac{d}{2}-1} e^{-\rho_1} \left\langle x_1 \left| e^{\frac{t}{4\rho_1}\Box} \right| z \right\rangle \int\limits_0^\infty d\rho_2 \rho_2^{\frac{d}{2}-1} e^{-\rho_2}$$
$$\times \left\langle x_2 \left| e^{\frac{t}{4\rho_2}\Box} \right| z \right\rangle \left[-\frac{C_1^{ab}}{\Gamma(d/2+1)} t\epsilon_1 \cdot \epsilon_2 + \frac{C_2^{ab}}{\Gamma(d/2)} \epsilon_1 \cdot (z-x_2)\, \epsilon_2 \cdot (z-x_1) \right]. \tag{48}$$

Here we have taken into account $\rho_1+\rho_2=\rho$ in the exponent and have included integrals over α into constants $C_{1,2}^{ab}$[2]:

$$C_1^{ab} = Tr\left(T^a T^b\right) 4^{d/2-1} C' \int\limits_0^1 d\alpha \ \left[\alpha(1-\alpha)\right]^{d/2-1} (\delta(\alpha)-1),$$
$$C_2^{ab} = Tr\left(T^a T^b\right) 4^{d/2-2} C' \int\limits_0^1 d\alpha \ \left[\alpha(1-\alpha)\right]^{d/2-2} (1-2\alpha)^2,$$
$$C' = g^2 \mu^{4-d} 2^{d/2-6} (2\pi)^{d/2-4}$$

We separate the integrals over the ρ_i, contained in heat kernel, and denote them as the $K_1(x_i,z,t)$ function:

$$K_1(x_i,z,t) = \int\limits_0^\infty d\rho_i \rho_i^{d/2-1} e^{-\rho_i} \left\langle x_i \left| e^{\frac{t}{4\rho_i}\Box} \right| z \right\rangle. \tag{49}$$

Identifying the variable t with the radius of d-dimensional sphere z_0 $\left(t = z_0^2\right)$ it can be shown that the function $K_1(x_i;z,t)$ obeys the $d+1$-dimensional massless Klein-Gordon equation:

$$\left[-z_0^2 \partial_{z_0}^2 + (d-1) z_0 \partial_{z_0} - z_0^2 \Box\right] K_1(x;z,t) = 0, \tag{50}$$

where $\Box$ is the d-dimensional Laplacian in the direction $\vec{z}$. We shall see that the function $K_1(x_i,z,t)$ is the bulk to boundary propagator of massless vector field in the $d+1$-dimensional AdS spacetime. Now we can write (48) in terms of this propagator:

$$\Pi^{ab}(x_1,x_2) = \int\limits_0^\infty \frac{dt}{t^{d/2+1}} \int\limits_{-\infty}^\infty d^d z \ K_1(x_1;z,t)\, K_1(x_2;z,t)$$
$$\times \left[-\frac{C_1^{ab}}{\Gamma(d/2+1)} \epsilon_1 \cdot \epsilon_2 t + C_2^{ab} \frac{1}{\Gamma(d/2)} \epsilon_1 \cdot (z-x_2)\, \epsilon_2 \cdot (z-x_1) \right]. \tag{51}$$

We note that the square bracket expansion in this case, differs from that of the free scalar theory by the t^3 factor. Thus, scalar loop contribution to

[2] We suppose $d \geq 2$ for convergence of these integrals. Finally, we put $d=4$.

the gluon polarization operator is expressed in terms of bulk to boundary propagator $K_1(x; z, t)$ of massless field.

For vector-spinor interaction case we have spinor particles in the loop and the two-point function is obtained from the one loop spinor effective action. Setting $N = 2$ in (23) and taking integrals over the Grassman variables, it reduces to the form of polarization operator found in the first lecture:

$$\Pi^{ab}(k_1, k_2) = 2Tr\left(T^a T^b\right) \frac{\left(g\mu^{\epsilon/2}\right)^2}{(4\pi)^{2-\epsilon/2}} \int_0^\infty \frac{d\tau}{\tau^{d/2-1}} \int_0^1 d\alpha \, e^{\tau k_1 k_2 \alpha(1-\alpha)} \int_{-\infty}^{\infty} \frac{d^d z}{(2\pi)^d} e^{i(k_1+k_2)z}$$
$$\times \left[\frac{2}{\tau}(\delta(\alpha)-1)\,\epsilon_1 \cdot \epsilon_2 + \left[(1-2\alpha)^2 - 1\right](\epsilon_1 \cdot k_2)(\epsilon_2 \cdot k_1) + (\epsilon_1 \cdot \epsilon_2)(k_1 \cdot k_2)\right]. \tag{52}$$

The worldline expression of two-point correlator for this interaction has minor difference with the scalar loop case in square bracket in (52). This enables us to re-express the spinor loop case in (52) similar to (51) as:

$$\Pi^{ab}(x_1, x_2) = 2\int_0^\infty \frac{dt}{t^{d/2+1}} \int_{-\infty} d^d z \, K_1(x_1; z, t)\, K_1(x_2; z, t) \left[C_1^{ab} \frac{1}{\Gamma(d/2+1)} \epsilon_1 \cdot \epsilon_2 t\right.$$
$$\left. + \frac{1}{\Gamma(d/2)} \left[C_2^{ab} \epsilon_1 \cdot (z - x_2)\, \epsilon_2 \cdot (z - x_1) - C_3^{ab} \epsilon_1 \cdot \epsilon_2 (z - x_1) \cdot (z - x_2)\right]\right]. \tag{53}$$

$$C_3^{ab} = Tr\left(T^a T^b\right) C' \int_0^1 d\alpha \; [\alpha(1-\alpha)]^{d/2-2}$$

Thus, we see from (53) that in AdS space the two-point correlator of massless vector field interacting with the spinor field is expressed by means of massless bulk to boundary propagator as well. We add scalar loop contribution to the fermion twice because of supersymmetry:

$$\Pi^{ab}(x_1, x_2) = 4C_2 \frac{1}{\Gamma(d/2)} \epsilon_1^i \epsilon_2^j \int_0^\infty \frac{dt}{t^{d/2+1}} \int d^d z \, K_1(x_1)\, K_1(x_2)$$
$$\times \left[(z - x_1)_i (z - x_2)_j - \delta_{ij} (z - x_1) \cdot (z - x_2)\right] \tag{54}$$

This should be comparable with the AdS two-point function result in (43). To this end we rewrite the supergravity correlator (43) in terms of $K_1(x_i)$ function. Taking derivatives from (41), using (49) and changing variable $z_0^2 = t$, we get the following expression for the two-point correlator (43):

$$\left\langle J_a^i(x_1)\, J_b^j(x_2)\right\rangle = \frac{1}{4g^2} \delta_{ab} \int_0^\infty \frac{dt}{t^{d/2+1}} t^{2-d/2} \int d^d z \, K_1(x_1)\, K_1(x_2)$$
$$\times \left[(z - x_1)_i (z - x_2)_j - \delta_{ij} (z - x_1) \cdot (z - x_2)\right]. \tag{55}$$

Comparing (55) and (54), we find out that the two-point super Yang-Mills correlator in the boundary of AdS spacetime has the same structure as that which comes from the supergravity in the bulk with $d = 4$.

Acknowledgements

Author thanks organizers for the opportunity to give these lectures in ISSSMB-06 in Mugla (Turkey) and hospitality.

References

1. M. J. Strassler, Nucl. Phys. B **385**, 145 (1992) [arXiv:hep-ph/9205205].
2. C. Schubert, Phys. Rept. **355**, 73 (2001) [arXiv:hep-th/0101036].
3. C. Schubert, "An Introduction to the Worldline Technique for Quantum Field Theory Calculations" [arXiv: hep-th/9610108]
4. Y. Ohnuki, T. Kashiwa, Prog. Theor. Phys. vol. 60, No 2 (1978), 548 Progress of Theoretical Physics Vol. 60 No. 2 (1978) pp. 548–564 URL: http://ptp.ipap.jp/link?PTP/60/548/ DOI : 10.1143/PTP.60.548
5. A.M. Polyakov, "Gauge fields and Strings", (Harwood Academic Publishers, 1987)
6. R. Gopakumar, Phys. Rev. D **70**, 025009 (2004) [arXiv:hep-th/0308184].
7. G. Chalmers, H. Nastase, K. Schalm and R. Siebelink, Nucl. Phys. B **540**, 247 (1999) [arXiv:hep-th/9805105].
8. Sh. Mamedov and S. Parvizi, "AdS/CFT correspondence via R-Current correlation functions revisited" [arXiv:hep-th/0505162]
9. Sh. Mamedov, "AdS interpretation of two-point correlation function of QED", Proceedings of 11-th Regional Conference on Mathematical Physics, 3–6 May, (Tehran, Iran) World Scientific, 2005, pp. 183–185, [arXiv:hep-th/0510006]
10. R.R. Metsayev and A.A. Tseytlin, Nucl. Phys. B **298**, 109 (1988)
11. Z.Bern, D.A. Kosower, Nucl. Phys. B **379**, 451 (1992) The computation of loop amplitudes in gauge theories Nuclear Physics B, Volume 379, Issue 3, 27 July 1992, Pages 451–561
12. F. Bastianelli and A. Zirotti, Nucl. Phys. B **642**, 372 (2002) [hep-th/0205182].

Course on Grand Unification

Goran Senjanović

ICTP, Trieste, Italy goran@ictp.it

1 Foreword

Although this is a field with a long history and a number of excellent reviews I tried to make this self contained, thus the section devoted to the group theory of SU(N) and SO(2N). For a newcomer in the field, this is must in my opinion. The material on spinors in SO(2N) is especially useful, and simple examples of SO(2), SO(4) and SO(6) are given in some detail. This should enable the reader to follow the physics and mathematics of spinors in SO(10) later on. The sections on SU(5) and Left-Right symmetry are fairly pedagogical and the reader armed with the knowledge of the first section should have no problems in following them. The last section devoted to the SO(10) theory, probably the most motived grand unified theory starts out being pedagogical, but the part which addresses the phenomenology of the ordinary and supersymmetric version is somewhat more difficult. If found too hard, it should be skipped in the first reading. It could be useful though for those interested in fermion masses and mixings. Throughout the text, I have included some exercises which I recommend.

Throughout these lectures I emphasize the issue of neutrino masses and mixings which the theory should reproduce and hopefully predict the ones not yet measured. Neutrino masses and mixings were discussed at this school by Henry Wong and I refer the reader to his lectures. A particularly useful review can be found in [1], while a pedagogical expose is provided in the book by Mohapatra and Pal [2]. Regarding grand unification, I recommend the books by Mohapatra [3] and Ross [4].

It is my hope that after going through this some of you will join the field and help in the search for the right theory.

2 Introduction and Basic Group Theory

The Standard Model (SM) of electro-weak and strong interactions is a remarkably successful theory of all particle forces but gravity. In its minimal version neutrinos are massless, but the observed tiny neutrino masses are easily accounted for through a new, high-energy physics; all it requires is to, say, add right-handed neutrinos, the SM singlets. True, one should find the Higgs particle in order to complete the theory, but to most of us it is only a question of time, rather likely to happen at LHC. Since it works so well, most of the attempts in building theories beyond the SM have focused on purely theoretical and even philosophical questions. One issue that stands out in my opinion is the disparity of three different forces based on SU(3), SU(2) and U(1) gauge groups. Particularly worrisome is the U(1), since its charge is not quantized, and the miracle of charge quantization in nature is accounted for by arbitrary of U(1) quantum numbers. This would be so if we had a single (or a product) non-abelian group, such as was the hope in the original proposal by Schwinger [5] of SU(2) as a unified theory of weak and electromagnetic interactions, long before the neutral currents were discovered. This guarantees charge quantization and is a prototype of any unified theory based on a simple gauge group The neutral gauge boson of SU(2) was to be identified with the photon and one had

$$Q = T_3 \tag{1}$$

Thus the charges were automatically, but unfortunately wrongly, quantized (a heroic attempt to save the theory was made by Georgi and Glashow [6], before the discovery of neutral currents). This could have been considered as an example of a beautiful theory killed by the ugly facts of nature. For charge quantization in this theory is a profound and deep fact: if one breaks SU(2) down to $U(1)_{em}$ as spontaneously as renormalizability requires, one predicts the existence of magnetic monopoles [7, 8]. This means that all the charges must be quantized, not just the ones of the observed particles. One is assured that the fact will persist whatever new particles be discovered.

Four years after Schwinger, Glashow made a simple but important suggestion: he added the U(1) piece [9]. The rest as we know is history. Well, what was missing was to break the symmetry through the Higgs mechanism which was to follow six years later [10,11], and everything fell in its place. It seemed quite a blow though that one was forced to introduce the U(1) culprit to get the lepton and quark charges right. Now, in the SM model the charges are quantized due to anomaly cancellation, but that does not say anything about the particles not yet discovered, their charges do not have to be quantized. For example, the vector-like states may have arbitrary real number charges since their anomalies cancel automatically.

But then came SU(3) as a theory of strong interactions, and a wish to unify both weak and strong forces in a simple theory, based on a single gauge group. This then, besides unification, leads to charge quantization automatically, and furthermore the minimal theory based on SU(5) gauge group [12], also includes U(1) for free. There are two generic beautiful predictions of grand unification: proton decay and magnetic monopoles, the former due to the unification of quarks and leptons and the latter due to in-built quantization of electric charge. Whereas Dirac predicted charge quantization if magnetic monopoles exist [13], grand unification predicts the existence of monopoles. Both proton decay and magnetic monopoles were searched for desperately, alas, with no success. Furthermore, the minimal grand unified theory of both matter and interactions based on SO(10) gauge group predicts massive neutrinos, and connects neutrino masses and mixings with the ones of the quarks. Thus this seems to be the right time to study grand unification more deeply, since its predictions may be testable in near future.

These lectures are devoted to the pedagogical expose of the idea of grand unification as well as up-to-date overview of new results. I focus on SU(5) and SO(10) theories with an in-depth attempt to create a minimal, standard theory, analogous to what was achieved in the electro-weak case. This is an extremely hard task due to the lack of experimental evidence, but a task worth pursuing. Since grand unification is its essence a rather ambitious extension of the SM, with no experimental motivation, I will try to stick to the basic issues and stay away from the complicated extensions of what is already quite a speculation. I will thus cover only models in four dimensions, and only those without any extra symmetries beyond the SU(5) or SO(10) symmetry. The only exception is supersymmetry: I include it since it is intimately related to unification through its deep connection to the hierarchy problem and since it was prophetic about the gauge coupling unification [14–17] through the large mass of the top quark [17]. On top of that, it can lead nicely to the tachyonic property of the Higgs, again through the large top quark mass [18, 19].

My aim is to have this review as self contained and as pedagogical as possible. In order to facilitate the discussion of physics in the following sections, I give here some basic discussion of tensor representations of SU(N) groups (very brief) and more interestingly of spinors in SO(2N) groups [20]. The latter are real fun to play with and that makes SO(10) theory, where each family of fermions fits into a 16 dimensional spinorial representation, quite interesting not only from the point of view of neutrino masses and mixings, but also from the point of view of group theory.

2.1 $SU(N)$

On a fundamental N-dimensional complex representation Φ, the $SU(N)$ group acts as

$$\Phi \to U\Phi, \quad U^\dagger U = 1, \quad det(U) = 1 \tag{2}$$

and U can be written as

$$U = e^{-i\theta_a T_a} \quad a = 1..N^2 - 1 \tag{3}$$

where the group generators T_a satisfy

$$T_a = T_a^\dagger, \quad Tr(T_a) = 0, \quad [T_a, T_b] = i f_{abc} T_c \tag{4}$$

where f_{abc} are the group structure constants. There is also a complex conjugate representation

$$\Phi^* \to U^* \Phi^* \tag{5}$$

and an $(N^2 - 1)$-dimensional adjoint representation

$$A \to U A U^\dagger = A - i\theta^a [T_a, A] + ... \tag{6}$$

In other words, the generators act on A as commutators. One can write $A = A_a T_a$, so that A_a transforms under a small group rotation as

$$A_a \to A_a + f_{abc} \theta_b A_c \tag{7}$$

Examples of fields transforming as the adjoint representation are the gauge bosons A of $SU(N)$ and the heavy scalars Σ employed to break the grand unified symmetry. The reason for the latter is the fact that under a unitary transformation $\langle \Sigma \rangle \to U \langle \Sigma \rangle U^\dagger$, one can have $\langle \Sigma \rangle$ diagonal, which in turn implies

$$[\langle \Sigma \rangle, T_a \in Cartan] = 0 \tag{8}$$

The adjoint Higgs preserves the rank of the group after the symmetry breaking. This is specially important in $SU(5)$ since it has the same rank ($= 4$) as the SM gauge group.

All other representations are built out of the fundamental Φ (and/or Φ^*) by symmetrizing and antisymmetrizing (and subtracting the trace when necessary). For example

$$\Phi_i \Phi_j = \Phi_{[i,j]} + \Phi_{\{i,j\}} \tag{9}$$

$$\frac{N(N-1)}{2} \quad \frac{N(N+1)}{2} \tag{10}$$

This means that all the charges get summed up

$$Q(\Phi_i \Phi_j) = Q(\Phi_i) + Q(\Phi_j) \tag{11}$$

2.2 $SO(2N)$

$SO(2N)$ is the group of real orthogonal transformations, $O^T O = O O^T = 1$, with $det(O) = 1$. It can be generated by $N(N-1)/2$ Hermitian antisymmetric matrices

$$O = e^{-i\theta_{ij} L_{ij}} \tag{12}$$

with

$$(L_{ij})_{kl} = -i(\delta_{ik}\delta_{jl} - \delta_{il}\delta_{jk}) \tag{13}$$

so that one has the following commutation relations

$$[L_{ij}, L_{kl}] = i(\delta_{ik}L_{jl} - \delta_{jl}L_{ik}) \tag{14}$$

The N-dimensional Cartan subalgebra is spanned by

$$Cartan = \{L_{12}, L_{34}, ..., L_{2N-1,2N}\} \tag{15}$$

whose eigenvalues are ± 1. The fundamental (vector) representation transforms as

$$\Phi_i \to O_{ij}\Phi_j \tag{16}$$

and is generated by L_{ij} in 13. One can construct the general N-index irreducible representation by antisymmetrizing or symmetrizing (and subtracting traces) N times the vector representation. Rather interesting are the [N]-index antisymmetric ones, for one can complexify them by introducing

$$\Phi^{\pm}_{[a_1..a_N]} = \Phi_{[a_1..a_N]} \pm \frac{i^N}{N!}\epsilon_{a_1...a_N b_1...b_N}\Phi_{b_1...b_N} \tag{17}$$

We illustrate this on a simple example below in SO(2) where this amounts to just complexifying a fundamental representation. It turns out that such 5 index antisymmetric 126 dimensional representation of SO(10) plays a profound role in a physics of neutrino mass; this is discussed in the last section.

SO(2*N*): spinors

By analogy with the Dirac algebra in Minkowski space, an Euclidean version is based on the Clifford algebra of the Γ_i matrices ($i = 1...2N$)

$$\{\Gamma_i, \Gamma_j\} = 2\delta_{ij} \tag{18}$$

out of which one can construct $N(N-1)/2$ generators

$$\Sigma_{ij} = \frac{1}{4i}[\Gamma_i, \Gamma_j] \tag{19}$$

which satisfy the usual commutation relations of the $SO(2N)$ generators in 14. It is easy to see that the Cartan subalgebra consists of N generators

$$Cartan = \{\Sigma_{12}, ..., \Sigma_{2N-1,2N}\} \tag{20}$$

whose eigenvalues are $\pm 1/2$.

The appropriate 2^N-dimensional complex representation Ψ is called a spinor of $SO(2N)$. Adding a spinor changes of course a group, just as SO(3) becomes SU(2). One often calls SO(10) with spinors Spin(2N). The spinors transforms in the following manner

$$\Psi \to e^{-i\theta_{ij}\Sigma_{ij}}\Psi \tag{21}$$

Again, by analogy with Dirac γ_5 matrix one can introduce

$$\Gamma_{\mathrm{FIVE}} = (-1)^N \Gamma_1 ... \Gamma_{2N} \tag{22}$$

with the properties

$$\Gamma_{\mathrm{FIVE}}^2 = 1, \quad [\Gamma_{\mathrm{FIVE}}, \Sigma_{ij}] = 0, \quad \{\Gamma_{\mathrm{FIVE}}, \Gamma_i\} = 0 \tag{23}$$

By using the projectors

$$\Gamma_{+(-)} \equiv \frac{1 \pm \Gamma_{\mathrm{FIVE}}}{2} \tag{24}$$

one can construct the irreducible 2^{N-1} dimensional spinors

$$\Psi_{\pm} \equiv \Gamma_{+(-)} \Psi \tag{25}$$

by analogy with Weyl spinors of the Lorentz group.

One can also introduce the analogue of the usual charge conjugation by demanding that

$$\Psi^T B \Psi = \text{invariant} \Leftrightarrow \Psi^c \equiv B \Psi^* \tag{26}$$

which amounts to

$$\Sigma^T B + B \Sigma = 0 \tag{27}$$

There are two possible solutions for B

$$B_{(1)} = \Gamma_1 ... \Gamma_{2N-1} \ , \ ; B_{(2)} = \Gamma_2 ... \Gamma_{2N} \tag{28}$$

The Ket Notation for Spinors

From

$$\Gamma_{\mathrm{FIVE}} = 2\Sigma_{12} .. 2\Sigma_{2N+1,2N} \tag{29}$$

one can write

$$\Gamma_{\mathrm{FIVE}} = \epsilon_1 \epsilon_2 ... \epsilon_N \tag{30}$$

where ϵ_i are ± 1, the eigenvalues of $\Sigma_{2i-1,2i}$. Then one can denote the Ψ_+ spinors as a ket

$$\Psi_+ \equiv |\epsilon_1 ... \epsilon_N\rangle \tag{31}$$

For example, take the spinors Ψ_+ of $SO(10)$

$$\Psi_+ \equiv |\epsilon_1 ... \epsilon_5\rangle; \quad \epsilon_1 .. \epsilon_5 = +1 \tag{32}$$

The 16-component Ψ_+ can be decomposed as

$$\Psi_+ = \begin{cases} \text{1 field} & |+++++\rangle \\ \\ \text{10 fields} & \begin{array}{l} |+++--\rangle, |++-+-\rangle, |++--+\rangle \\ |+-++-\rangle, |+-+-+\rangle, |+--++\rangle \\ |-+++-\rangle, |-++-+\rangle, |-+-++\rangle, |--+++\rangle \end{array} \\ \\ \text{5 fields} & \begin{array}{l} |+----\rangle, |-+---\rangle \\ |--+-\rangle, |---+-\rangle, |----+\rangle \end{array} \end{cases} \tag{33}$$

We will see that this can be interpreted as a decomposition under $SU(5)$

$$16 = 10 + \overline{5} + 1 \tag{34}$$

In other words, a family of fermions augmented by a right-handed neutrino makes and irreducible spinorial representation of $SO(10)$. The unification of matter, on top of gauge interactions, points strongly towards $SO(10)$. However, in order to appreciate this fact and have fun with $SO(10)$, we first go through some pedagogical exposition of smaller groups.

$SO(2)$: A Prototype for $SO(4n+2)$

We choose

$$\Gamma_1 = \sigma_1,\ \Gamma_2 = \sigma_2 \tag{35}$$

so that

$$\Gamma_{\text{FIVE}} = \sigma_3 \quad \Sigma_{12} = \frac{\sigma_3}{2} \tag{36}$$

which illustrates clearly $[\Gamma_{\text{FIVE}}, \Sigma_{i,j}] = 0$. The irreducible 1-component spinors transform as

$$\Psi_+ \to e^{-i\theta/2}\Psi_+\,, \quad \Psi_- \to e^{+i\theta/2}\Psi_- \tag{37}$$

since

$$\Psi \equiv \begin{pmatrix} \Psi_+ \\ \Psi_- \end{pmatrix} \to e^{-i\theta\sigma_3/2} \begin{pmatrix} \Psi_+ \\ \Psi_- \end{pmatrix} \tag{38}$$

On the other hand, the two-component vectors transform as

$$\begin{pmatrix} \phi_1 \\ \phi_2 \end{pmatrix} \to \begin{pmatrix} \cos\theta & \sin\theta \\ -\sin\theta & \cos\theta \end{pmatrix} \begin{pmatrix} \phi_1 \\ \phi_2 \end{pmatrix} \tag{39}$$

or

$$\phi_1 \pm i\phi_2 \to e^{\pm i\theta}(\phi_1 \pm i\phi_2) \tag{40}$$

Eqs. 37 and 40 simply account for the fact $SO(2) \simeq U(1)$.

The internal "charge" conjugation B can be chosen as $B_1 = \sigma_1$, so that

$$\Psi^T B \Psi = \Psi_+ \Psi_- \tag{41}$$

However, only Ψ_+ (or Ψ_-) is an irreducible spinor, therefore there is no mass term for an irreducible spinor of $SO(2)$. In other words, the spinors Ψ_+ (Ψ_-) are chiral and can represent physical particles such as the fermions of the SM. This is true in any $SO(4n+2)$ theory. In particular, in $SO(10)$, which means that it offers hope of being realistic.

Dual representation

From

$$\epsilon_{ij} det\, O = O_{ik} O_{jl} \epsilon_{kl} \tag{42}$$

is easy to see that ϕ_i and $\epsilon_{ij}\phi_i$ transform in the same way. We can introduce the self (anti-self) dual representation

$$\Phi_i(\pm) = \frac{1}{\sqrt{2}}(\phi_i \pm i\epsilon_{ij}\phi_j) \tag{43}$$

which is nothing else but the complex representation of U(1) 40. This should make clear the generic concept of self dual representations in $SO(2N)$ discussed before.

Yukawa couplings

We have seen that there is no direct mass term. There are Yukawa couplings, though, of the type

$$\begin{aligned}\mathcal{L}_Y &= \Psi^T B \sigma_i \Psi \phi_i \\ &= \Psi_+ \Psi_+ (\phi_1 - i\phi_2) + \Psi_- \Psi_- (\phi_1 + i\phi_2)\end{aligned} \tag{44}$$

as dictated by $U(1)$ charges.

$SO(4)$

One knows that $SO(4)$ is isomorphic to $SU(2) \times SU(2)$, and it plays an important role in providing a left-right symmetric subgroup of $SO(10)$. It is an Euclidean analog of the Lorentz group and the Clifford algebra can be generated by

$$\begin{aligned}&\Gamma_1 = \begin{pmatrix} 0 & \sigma_1 \\ \sigma_1 & 0 \end{pmatrix} \; \Gamma_2 = \begin{pmatrix} 0 & \sigma_2 \\ \sigma_2 & 0 \end{pmatrix} \\ &\Gamma_3 = \begin{pmatrix} 0 & \sigma_3 \\ \sigma_3 & 0 \end{pmatrix} \; \Gamma_4 = \begin{pmatrix} 0 & -i \\ i & 0 \end{pmatrix}\end{aligned} \tag{45}$$

so that

$$\Gamma_{\text{FIVE}} = \begin{pmatrix} 1 & 0 \\ 0 & -1 \end{pmatrix} \tag{46}$$

and "charge" conjugation can be taken as

$$B_{(1)} = \Gamma_1 \Gamma_3 = \begin{pmatrix} -i\sigma_2 & 0 \\ 0 & -i\sigma_2 \end{pmatrix} \tag{47}$$

or

$$B_{(2)} = \Gamma_2 \Gamma_4 = \begin{pmatrix} i\sigma_2 & 0 \\ 0 & -i\sigma_2 \end{pmatrix} \tag{48}$$

The mass term

$$\Psi^T B \Psi \propto \Psi_+^T i\sigma_2 \Psi_+ + ... \tag{49}$$

where

$$\Psi_\pm = \frac{1 \pm \Gamma_5}{2} \Psi_\pm \tag{50}$$

In other words, the mass term for Ψ_+ (or Ψ_-) is invariant, which means that we can have no chiral fermions in $SO(4)$. This is true for all $SO(4n)$ groups.

In the ket notation

$$\Psi_+ = |\epsilon_1 \epsilon_2\rangle; \quad \epsilon_1 \epsilon_2 = 1; \quad \epsilon_{1,2} = \pm 1 \tag{51}$$

or

$$\Psi_+ = \begin{pmatrix} |++\rangle \\ |--\rangle \end{pmatrix} \tag{52}$$

Introduce the neutral generator of $SU(2)_L$ and $SU(2)_R$

$$T_{3L} \equiv \frac{1}{2}(\Sigma_{12} + \Sigma_{34}) \, , \quad T_{3R} \equiv \frac{1}{2}(\Sigma_{12} - \Sigma_{34}) \tag{53}$$

and you see that Ψ_+ is an $SU(2)_L$ doublet, $SU(2)_R$ singlet field, an analog of left-handed Weyl spinors of the Lorentz group. Similarly, Ψ_- is an $SU(2)_L$ singlet, $SU(2)_R$ doublet field.

SO(6)

$SO(6) \sim SU(4)_C$ is the Pati-Salam group of quark-lepton symmetry, with leptons as the fourth color. It deserves a brief description.

Start with a six-dimensional vector Φ_i (i=1..6). It is easy to see that the components $(\phi_1 \pm \phi_2)$, $(\phi_3 \pm \phi_4)$, $(\phi_5 \pm \phi_6)$ transform as 3 and 3* of its subgroup $SU(3)$ which we identify with the color.

The neutral generators are identified as

$$\begin{aligned} T_{3C} &= \frac{1}{2}(\Sigma_{12} - \Sigma_{34}) \\ T_{8C} &= \frac{1}{2}(\Sigma_{12} + \Sigma_{34} - 2\Sigma_{56}) \end{aligned} \tag{54}$$

The additional neutral generator of $SU(4)$, identifiable as $B-L$, can be written as

$$B - L = -\frac{2}{3}(\Sigma_{12} + \Sigma_{34} + \Sigma_{56}) \tag{55}$$

Regarding spinors, the positive chirality can be written as

$$\Psi_+ = \begin{cases} \text{color singlet} & |+++\rangle \\ \text{color triplet (B} - \text{L)} = 1/3 & |+--\rangle, |-+-\rangle, |--+\rangle \end{cases} \tag{56}$$

It says simply that the irreducible 4-component spinor of $SO(6)$ is a fundamental of $SU(4)$ with the decomposition under $SU(3)_C$ (with $B-L$)

$$\Psi_+ = 4 = 1_{-1} + 3_{1/3} \tag{57}$$

which is precisely a combination of a lepton and a colored quark. Similarly, $\Psi_i = 4^* = 1_{+1} + 3_{-1/3}$ stands for an antilepton and antiquark.

Exercise 1 *As a check, show that* $4 \times 4 = 6 + 10$. *Show that* 6 *of* $SO(6)$ *has the quantum number of the* 6 *(antisymmetric) of* $SU(4)$.

Yukawa couplings in $SO(6)$

We know that the irreducible spinors of $SO(6)$ are fundamental representations of $SU(4)$ and $4 \times 4 = 6 + 10$. There are then two types of Yukawa couplings

$$\mathcal{L}_Y = y_6 \Psi^T B \Gamma_i \Psi \Phi_i + y_{10} \Psi^T B \Gamma_i \Gamma_j \Gamma_k \Psi \Phi^-_{[ijk]} \tag{58}$$

where it is a simple exercise to show that $\Phi^-_{[ijk]}$is an anti-self-dual representation

$$\Phi^-_{ijk} = \Phi^-_{[ijk]} = \frac{i}{3!} \epsilon_{ijklmn} \Phi_{[lmn]} \tag{59}$$

and where $\Phi^-_{[ijk]}$ is the 3-index antisymmetric tensor of $SO(6)$.

Exercise 2 *Construct the self-dual and anti-self-dual representation of* $SO(6)$ *out of the 3-index antisymmetric representation* $\Phi_{[ijk]}$. *Show that* $20 = 10 + \overline{10}$. *Then prove equation 58 and show that there are no other couplings.*

Exercise 3 *Take the Pati-Salam group* $SO(4) \times SO(6) \simeq SU(2)_L \times SU(2)_R \times SU(4)_c$. *Show that the representations* $(2, 1, 4)$ *and* $(1, 2, \overline{4})$ *give a family of quarks and leptons augmented by a right-handed neutrino.*

Exercise 4 *The chiral anomalies are proportional to* $\Lambda_{ijk} = Tr(\{T_i, T_j\}T_k)$. *Show that the* $SO(2N)$ *groups are anomaly free, except for the* $SO(6)$. *Comment on why* $SO(6)$ *must have an anomaly.*

3 SU(5): A Prototype GUT

The minimal group that can unify the Standard Model (SM) is $SU(5)$, a group of rank four. It is actually the minimal group that can unify the $SU(2)_L$ and $SU(3)_c$ of the SM, the $U(1)$ comes for free.

It is natural that we should try to put the electro-weak doublet Φ and the new color triplet h_α in the 5-dimensional fundamental representation

$$5_H = \Phi = \begin{pmatrix} h^r \\ h^g \\ h^b \\ \phi^+ \\ \phi^0 \end{pmatrix} \begin{matrix} \left.\begin{matrix} \\ \\ \\ \end{matrix}\right\} SU(3)_c \\ \left.\begin{matrix} \\ \\ \end{matrix}\right\} SU(2)_L \end{matrix} \tag{60}$$

where in the obvious notation the $SU(3)_c$ symmetry is acting on the first 3 components and the $SU(2)_L$ on the last two.

3.1 Structure

Fermions

We have 15 Weyl fields in each generation and it is natural to try to put them in a 15-dimensional symmetric representation of $SU(5)$. Now

$$5 \otimes 5 = 15_s + 10_{as} \tag{61}$$

Since $5 = (3_c\,, 1_L) + (1_c\,, 2_L)$ (in an obvious notation), since $(3_c \otimes 3_c)_s = 6_c$, and since quarks come only in color triplets, we must abandon the idea of 15_S. It is not anomaly free anyway, it could not have worked. What about 5 and 10_{as} ? The quantum numbers of 5 from (60) imply uniquely

$$5_F \equiv \psi = \begin{pmatrix} d^r \\ d^g \\ d^b \\ e^+ \\ -\nu^C \end{pmatrix}_R \tag{62}$$

(recall that $(f^C)_R \equiv C\bar{f}_L$).

Now, from $\psi \longrightarrow U\psi$ under $SU(5)$, the 10-dimensional representation χ must transform as

$$\chi \longrightarrow U\,\chi\,U^T \tag{63}$$

This is enough to give the quantum numbers of the particles in 10

$$\chi = \frac{1}{\sqrt{2}} \begin{bmatrix} 0 & u_b^C & -u_g^C & -u^r & -d^r \\ -u_b^C & 0 & u_r^C & -u^g & -d^g \\ u_g^C & -u_r^C & 0 & -u^b & -d^b \\ u^r & u^g & u^b & 0 & e^+ \\ d^r & d^g & d^b & -e^+ & 0 \end{bmatrix}_L \tag{64}$$

Notice that in (62), a minus sign convention for the ν^C field is to ensure that $\begin{pmatrix} e^+ \\ -\nu^C \end{pmatrix}_R$ and $\begin{pmatrix} e \\ \nu \end{pmatrix}_L$ transform identically, and in (64) the signs are the property of χ being antisymmetric. We will work in the future with 10_F and $\bar{5}_F$ (instead of 5_F).

We can see furthermore that a unified theory such as $SU(5)$ explains charge quantization, i.e. it relates quark and lepton charges. From (62)

$$Q(d^C) = -\frac{1}{3}Q(e) = \frac{1}{3} \tag{65}$$

and then from (64) we see that $Q(u) = Q(d) + 1 = 2/3$.

$SU(5)$: Interactions

The interactions of fermions with gauge bosons are

$$\mathcal{L}_f = i\bar{\psi}\gamma^\mu D_\mu \psi - iTr\bar{\chi}\gamma^\mu D_\mu \chi \tag{66}$$

where

$$D_\mu \chi = \partial_\mu \chi - ig(\mathcal{A}_\mu \chi + \chi \mathcal{A}_\mu^T) \tag{67}$$

There are of course the old QCD and $SU(2)_L \times U(1)$ interactions with $g_s = g_W = g$, and $sin^2\theta_W = 3/8$, the couplings at the unification scale where full SU(5) is operative. Furthermore, there are new X and Y bosons who carry both color and flavor with charges 4/3 and 1/3 respectively. Their interactions are

$$\begin{aligned}\mathcal{L}(X,Y) = & \frac{g}{\sqrt{2}}\bar{X}_\mu^\alpha \left[\bar{d}_{\alpha R}\gamma^\mu e_R^+ + \bar{d}_{\alpha L}\gamma^\mu e_L^+ + \epsilon_{\alpha\beta\gamma}\bar{u}_L^{c\gamma}\gamma_\mu u_{\beta L}\right] \\ & + \frac{g}{\sqrt{2}}\bar{Y}_\mu^\alpha \left[-\bar{d}_{\alpha R}\gamma^\mu \nu_R^C + \bar{u}_{\alpha L}\gamma^\mu e_L^+ + \epsilon_{\alpha\beta\gamma}\bar{u}_L^{c\gamma}\gamma_\mu d_{\beta L}\right] + \quad h.c.\end{aligned} \tag{68}$$

As expected, due to the nontrivial color and flavor characteristics of the quarks, the X and Y couple to the quark-quark and quark-lepton states. It is clear that B and L are violated, although for some magic reason $B - L$ is conserved (more about it later). This leads to the decay of the proton. By analogy with the usual weak decay $n \to p + e + \bar{\nu}$, $\mu \to e + \bar{\nu}_e + \nu_\mu$ the proton decay rate can be estimated as

$$\Gamma_p \simeq \frac{g^4}{M_X^4} m_p^5 \tag{69}$$

From $(\tau_p)_{exp} > 10^{33} yr$ we get $M_X > 10^{15.5} GeV$; later we will show that we can actually compute M_X.

3.2 Symmetry Breaking

The first stage of symmetry breaking down to the SM is achieved by the adjoint Higgs $\Sigma = 24_H$. Assume, only for the sake of simplicity, the discrete symmetry $\Sigma \to -\Sigma$. Then the most general renormalizable potential for Σ is given by

$$V(\Sigma) = -\frac{\mu^2}{2}\, Tr\Sigma^2 + \frac{1}{4}a\,(Tr\Sigma^2)^2 + \frac{1}{2}b\, Tr\Sigma^4 \tag{70}$$

Now, since $\langle\Sigma\rangle$ is a Hermitean matrix it can be diagonalized by an $SU(5)$ rotation. Assume now that it is in the same direction as the hypercharge: $\langle\Sigma\rangle \propto Y = v_X\, diag(1,1,1,-3/2,-3/2)$.

From (70) you get then $\mu^2 = \frac{1}{2}(15a+7b)\, v_X^2$, which, for $\mu^2 > 0$, implies $(15a+7b) > 0$.

In order to check that this is a local minimum, we must show that all the second derivatives are positive. Since Σ has exactly the same form as the gauge boson matrix, we can write

$$\Sigma = \langle\Sigma\rangle + \begin{pmatrix} \Sigma_8 + \sqrt{\frac{3}{5}}\left(-\frac{2}{3}\right) 1_c \Sigma_0 & \bar{\Sigma}_X & \bar{\Sigma}_Y \\ \Sigma_X & \sqrt{\frac{1}{2}}\Sigma_3 + \sqrt{\frac{3}{5}}\Sigma_0 & \Sigma^+ \\ \Sigma_Y & \Sigma^- & -\sqrt{\frac{1}{2}}\Sigma_3 + \sqrt{\frac{3}{5}}\Sigma_0 \end{pmatrix} \tag{71}$$

where Σ_8 are the analogs of gluons, Σ_X and Σ_Y the analogs of X and Y, Σ_3, Σ^+, Σ^- and Σ_0 the analogs of W^3, W^+, W^- and B, respectively. The masses of the particle masses in Σ are

$$\begin{aligned} m^2(\Sigma_8) &= \frac{5}{4} b\, v_X^2 m^2(\Sigma_3) = m^2(\Sigma_\pm) \;=\; 5b\, v_X^2 \\ m^2(\Sigma_0) &= \frac{15a+7b}{2}\, v_X^2 m^2(\Sigma_X) = m^2(\Sigma_Y) \;=\; 0 \end{aligned} \tag{72}$$

Thus for $15a+7b > 0, b > 0$ the extremum is a local minimum of the theory. Notice that Σ_X and Σ_Y are would-be Goldstone bosons of the theory; they get "eaten" by the X and Y gauge fields, i.e. they become their longitudinal components.

Finally, one can show that the vev of Σ is actually a global minimum. In fact, other extrema can be shown to be at best saddle points.

Exercise 5 *Prove that the above minimum is in fact global*

Thus $SU(5)$ can be successfully broken down to the standard model, since as we said Y commutes with both the $SU(3)_c$ and $SU(2)_L \times U(1)_Y$ generators. This will be even more evident from the study of the gauge bosons mass matrix. Since Σ is in the adjoint representation, $D_\mu \Sigma = \partial_\mu \Sigma - ig[\mathcal{A}_\mu, \Sigma]$, and one has

$$\frac{1}{2}(D_\mu < \Sigma >)^\dagger (D^\mu < \Sigma >) = \frac{25}{8} g^2 v_X^2 \left[\bar{X}^a_\mu X^\mu_a + \bar{Y}^a_\mu Y^\mu_a \right] \tag{73}$$

where a as usual is the color index, $a = r,\ g,\ b$. As expected, the gluons and the electro-weak gauge bosons remain massless, but X and Y get equal masses

$$m_X^2 = m_Y^2 \equiv M_X^2 = \frac{25}{8} g^2 v_X^2 \tag{74}$$

as a consequence of both $SU(3)_c$ and $SU(2)_L$ remaining unbroken. The original $SU(5)$ symmetry is broken down to $SU(3)_c \times SU(2)_L \times U(1)_Y$.

The rest of the breaking is completed by a 5-dimensional Higgs multiplet Φ_5 which contains the Standard Model doublet. Let us study this in some detail including the full $SU(5)$ invariant potential. We can write

$$\begin{aligned} V(\Sigma, \Phi) = &- \frac{\mu_\Sigma^2}{2}\, Tr\Sigma^2 + \frac{1}{4} a (\, Tr\Sigma^2)^2 + \frac{1}{2} b\, Tr\Sigma^4 \\ &- \frac{\mu_\Phi^2}{2} \Phi^\dagger \Phi + \frac{\lambda}{4} (\Phi^\dagger \Phi)^2 \\ &+ \alpha \Phi^\dagger \Phi\, Tr\Sigma^2 - \beta \Phi^\dagger \Sigma^2 \Phi \end{aligned} \tag{75}$$

with $a > 0$, $\lambda > 0$, $15a + 7b > 0$ and $\beta > 0$. Since both $SU(3)_c$ and $SU(2)_L$ are unbroken at this point, we can always rotate $\langle \Phi \rangle$ into the form $\langle \Phi^T \rangle = (v_c, 0, 0, 0, v_W)$. It is only the β term that is sensitive to the direction of $< \Phi >$ and it gives $-\beta v_X^2 (v_c^2 + 9/4 v_W^2)$, which for $\beta > 0$ forms the solution $v_W \neq 0$, $v_c = 0$ in order to minimize the energy.

It is an easy exercise to compute the mass of the colored triplet scalar h_a in $\langle \Phi \rangle$, it is $m_h^2 = \frac{5}{2} \beta v_X^2$, which justifies the choice $\beta > 0$. It is also easy to show that

$$M_W^2 = \frac{g^2}{4\lambda} \left[\mu_\Phi^2 + \frac{8 M_X^2}{25 g^2} (-15\alpha + \frac{9}{2}\beta) \right] \tag{76}$$

But $M_X > 10^{15} GeV$, which implies an extraordinary fine-tuning in the above equation of at least 26 orders of magnitude. The number on the right hand side of (76) is naturally of order $M_X^2 > 10^{30} GeV^2$; instead it ends up being $\simeq (100 GeV)^2$. This is known as *the hierarchy problem.*

In the next subsection we will see that the colored triplet h_a mediates proton decay and thus it must be very heavy: $m_h > 10^{12} GeV$, implying that β cannot be taken arbitrarily small. On the other hand, its partner η weighs <1 TeV, and this aspect of the hierarchy problem is known as the *doublet-triplet splitting problem.*

Before we close this subsection, let us say a few words more on the hierarchy problem. The problem is that the mass term for the Higgs scalars cannot be made small (or zero) by any symmetry, unlike the case of fermions. There the limit $m_f = 0$ corresponds to the chiral symmetry $f \to \gamma_5 f$, and thus the higher order corrections must also vanish if $m_f = 0$ at the tree level. In other words, the higher order corrections are necessarily proportional to $m_f(tree)$, and so only logarithmically divergent. In the case of scalars the divergence is quadratic and thus in the context of grand unified theories (GUTs) such as $SU(5)$ the natural value for M_W is of order M_X.

3.3 Yukawa Couplings and Fermion Masses

In the Standard Model the left-handed fermions are doublets and the right-handed fermions are singlets, and so their chiral property is more than manifest. In $SU(5)$ the V-A structure of a family of fermions is left-intact and here also there are no direct mass terms for fermions.

In the minimal $SU(5)$ theory the fermion masses originate through the Yukawa couplings of fermions with the light Higgs Φ

$$\mathcal{L}_Y = f_d \, \bar{\psi}_R \, \chi \, \Phi^\dagger + f_u \frac{1}{2} \chi^T \, C \, \chi \, \Phi + h.c. \tag{77}$$

where C is the Dirac conjugation matrix, and f_u is clearly a symmetric matrix. The symbolic notation of (77) should read in the $SU(5)$ notation as

$$\begin{aligned} \bar{\psi}_R \, \chi \, \Phi^\dagger &= \bar{\psi}_{R\,i} \, \chi^{ij} \, \Phi^\dagger_j \\ \chi^T \, C \, \chi \, \Phi &= \epsilon_{ijklm} \, (\chi^T)^{ij} \, C \, \chi^{kl} \, \Phi^m \end{aligned} \tag{78}$$

With $< \Phi >^T = (0000 v_W)$, we get for fermionic masses

$$\begin{aligned} \mathcal{L}_m &= f_d v_W (\bar{d}_R d_L + \bar{e}^+_R e^+_L) - f_u v_W (u^c)^T_L \, C \, u_L + h.c. \\ &= -[f_d v_W (\bar{d}d + \bar{e}e) - f_u v_W \bar{u}u] \end{aligned} \tag{79}$$

In other words, just as in the Standard Model $m_f = h_f v_W$, furthermore charged lepton and down quark masses are equal.

Exercise 6 *Explain why this happens*

Unfortunately, this works bad even for the third family, since atM_X one finds $m_b = 0.6 m_\tau$. This means that one must include higher dimensional operators in the Yukawa sector, up to now neglected. Alternatively, you can include other Higgs representations that can contribute to the fermionic masses; for example, you can add 45_H.

Now, besides the usual Yukawa structure of the Higgs doublet in the SM, one has new interactions of the color triplet h_α. From (77) and (78) it is easy to compute it's couplings to fermions

$$\mathcal{L}_h = f_d \bar{\psi}_{R\,i} \chi^{i\,\alpha} h^+_\alpha + f_u \epsilon_{ijkl\alpha} (\chi^T)^{ij} C \chi^{kl} h^\alpha \tag{80}$$

which gives

$$\mathcal{L}_h = \left\{ f_d \left(\epsilon^{\alpha\beta\gamma} \bar{u}^c_{L\,\beta}\, d^\gamma_R + \bar{u}^\alpha_L\, e^+_R + \bar{d}^\alpha_L\, \nu^c_R \right) \right.$$
$$\left. + f_u \left(\epsilon^{\alpha\beta\gamma} \bar{u}^c_{R\,\beta}\, d^\gamma_L + \bar{u}^\alpha_R\, e^+_L \right) \right\} h^\alpha \qquad (81)$$

Notice that the structure of the above couplings (not the strength, though), is dictated by the $SU(3)_C \times SU(2)_L \times U(1)_Y$ gauge invariance only. This becomes more clear if we write $\bar{u^c_L} d_R = u^T_R C d_R$ and $\bar{u^c_R} d_L = u^T_L C d_L$.

It is clear that the interactions of H break B and L, just like those of X and Y. Notice, though, that $B-L$ is again conserved. In a complete analogy with the situation encountered before for the X and Y bosons, we have the possible exchanges of h^α which leads to the proton decay. Of course, the amplitude is proportional to small Yukawa couplings and the corresponding limit on its mass is somewhat less strict: $m_h \geq 10^{12} GeV$.

Generations and Their Mixings

We know that in the standard model the neutral current interactions are flavor diagonal and that the charged current processes lead to flavor mixing and CP violation. How is this feature incorporated in the $SU(5)$ theory and what about new superweak interactions of the X and Y bosons? The analysis is straightforward and it proceeds along the same lines as in the $SU(2)_L \times U(1)_Y$ theory [21]. I should stress that the predictions we will obtain are of course not realistic since in this minimal theory neutrinos are massless and the down quark and charged lepton mass relations come out wrong. The minimal model discussed here should be viewed only as a prototype of the what predictive theory should be like.

We diagonalize as usual fermion mass matrices by bi-unitary transformations

$$\mathcal{U}^\dagger_{Lf} M_f \mathcal{U}_{Rf} = D_f \qquad (82)$$

where D_f is diagonal, with its elements being real, positive numbers. Furthermore, since M_u is symmetric

$$\mathcal{U}_{Ru} = \mathcal{U}^*_{Lu} K^* \qquad (83)$$

where

$$K = \begin{pmatrix} e^{i\phi u} & & & \\ & e^{i\phi e} & & \\ & & e^{i\phi t} & \\ & & & \cdots \end{pmatrix} \qquad (84)$$

is the matrix of phases needed to ensure that the elements of D_u are real and positive. The above statements are equivalent to the redefinition of our original fermionic fields in the Lagrangian

$$f_{L,R} \rightarrow \mathcal{U}^\dagger_{L,R} f_{L,R} \qquad (85)$$

with $\mathcal{U}^d_{L,R} = \mathcal{U}^{e^+}_{L,R}$. Since on the other hand the neutrinos are massless, we can rotate them any which way we wish and so we chose $\nu^c_R \to \mathcal{U}^d_R \nu^c_R$. Thus we can write for the 5-dimensional representation $\psi_R \to \mathcal{U}^d_R \psi_R$, which means that $\mathcal{U}^d_R$ disappears since it is just an overall factor. Suppressing the color index, we can write

$$\chi \to \begin{bmatrix} \mathcal{U}_{Lu} K u^c & -\mathcal{U}_{Lu} u & -\mathcal{U}_{Ld} d \\ & & \\ & & -\mathcal{U}_{Ld} e^+ \end{bmatrix}_L$$

$$= \mathcal{U}_{Ld} \begin{bmatrix} \mathcal{U}_{CKM} K u^c & -\mathcal{U}_{CKM} u & -d \\ & & \\ & & -e^+ \end{bmatrix}_L \tag{86}$$

where $\mathcal{U}_{CKM} = \mathcal{U}^\dagger_{Ld}\mathcal{U}_{Lu}$. Again $\mathcal{U}_{Ld}$ is just an overall factor and so it will disappear. We are left with the Cabibbo-Kobayashi-Maskawa unitary matrix and the phase matrix K only. Thus the leptonic interactions are flavor conserving (since neutrinos are massless), and the weak quark interactions involve $\mathcal{U}_{CKM}$ only, as it must be. Finally, the X and Y boson interactions involve *no* new flavor mixings besides $\mathcal{U}_{CKM}$, however there will be new phases hidden in K. In the physical basis we get

$$\begin{aligned} \mathcal{L}(X,Y) = & \frac{g}{\sqrt{2}} \bar{X}_\mu \left[\bar{d}_R \gamma^\mu e^+_R + \bar{d}_L \gamma^\mu e^+_L + \bar{u}^c_L \gamma^\mu K^* u_L \right] \\ & + \frac{g}{\sqrt{2}} \bar{Y}_\mu \left[-\bar{d}_R \gamma^\mu \nu^c_R + \bar{u}_L \gamma^\mu \mathcal{U}^\dagger_{CKM} e^+_L + \bar{u}^c_L \gamma^\mu \mathcal{U}^\dagger_{CKM} d_L \right] + h.c. \end{aligned} \tag{87}$$

From $\mathcal{U}_{11} \propto \cos\theta_c$, $\mathcal{U}_{12} \propto \sin\theta_c$ we would expect

$$\frac{\Gamma(p \to \pi^0 \mu^+)}{\Gamma(p \to \pi^0 e^+)} \propto \sin^2\theta_c \tag{88}$$

Of course, this minimal SU(5) model is not realistic, for down and strange quark masses are not equal to their leptonic counterparts at the unification scale. It is only an illustration how proton decay partial rates are connected to the fermion masses and mixings. The true test can only be possible in a completely realistic theory of fermion masses and mixings (for a review and references, see [22]).

In any case, the minimal SU(5) theory fails to explain neutrino masses; it is custom fit for massless neutrinos. While non-minimal models can lead to non-vanishing neutrino masses, by itself, SU(5) just like the standard model cannot relate neutrino masses to charged fermion masses nor relate quark and lepton mixing angles. This is cured beautifully in the SO(10) theory which requires the existence of right-handed neutrinos and leads to small, non-vanishing neutrino masses through the see-saw mechanism. The main ingredients are the left-right and quark-lepton symmetry inbuilt in SO(10) automatically. However, SU(5) offers an interesting possibility of neutrino Yukawa couplings be probed at LHC and before moving to SO(10) in the last section we will discuss a simple and predictive SU(5) theory with an adjoint fermionic representation added to the minimal model discussed above. We will show that the theory is completely realistic and testable at colliders.

3.4 Low Energy Predictions

Ordinary $SU(5)$

As is well known, the couplings run logarithmically with energy. We have

$$\frac{1}{\alpha_G(M_W)} = \frac{1}{\alpha_U} - \frac{1}{2\pi} b_G \ln \frac{M_X}{M_W} \tag{89}$$

for the gauge group G; M_X is the energy where we imagine the unification to take place, and α_U is the value of the unified coupling at M_X. One has a generic formula for the running coefficient

$$b_G = \frac{11}{3} T_{GB} - \frac{2}{3} T_F - \frac{1}{3} T_H \tag{90}$$

This gives for the $SU(3)_C$, $SU(2)_L$ and $U(1)$ respectively

$$\begin{aligned} b_3 &= \frac{33}{3} - \frac{4}{3} n_g \\ b_2 &= \frac{22}{3} - \frac{4}{3} n_g - \frac{1}{6} n_H \\ b_1 &= \frac{3}{5} b_Y = -\frac{4}{3} n_g - \frac{1}{10} n_H \end{aligned} \tag{91}$$

where N_g is the number of generations, n_H is the number of Higgs doublets ($n_H = 1$ in the minimal standard model).

We are now fully armed to check the evolution of these couplings above M_W. From above

$$\frac{1}{\alpha_i(M_W)} - \frac{1}{\alpha_j(M_W)} = \frac{b_j - b_i}{2\pi} \ln \frac{M_X}{M_W} \tag{92}$$

In the above we have used $\alpha_1(M_X) = \alpha_2(M_X) = \alpha_3(M_X) = \alpha_U$. From $\alpha_{em} = \sin^2\theta_W \alpha_2 = \cos^2\theta_W \alpha_Y$ and $\alpha_Y = 3/5\alpha_1$ we get easily

$$\frac{1}{\alpha_2(M_W)} - \frac{1}{\alpha_3(M_W)} = \frac{22 + n_H}{12\pi} \ln \frac{M_X}{M_W}$$

$$\sin^2\theta_W(M_W) = \frac{3}{8} - \frac{110 - n_H}{48\pi} \alpha_{em}(M_W) \ln \frac{M_X}{M_W} \tag{93}$$

Notice the prediction $\sin^2\theta_W = \frac{3}{8}$ at M_X which we discussed before. Now, for $n_H = 1$ and by taking a $\alpha_3(M_W) \simeq .12$, $\alpha_2(M_W) \simeq \frac{1}{30}$ we find $M_X \simeq 10^{16} GeV$, but

$$\sin^2\theta_W(M_W) \simeq 0.2 \tag{94}$$

The minimal $SU(5)$ theory thus fails to meet the experiment.

Supersymmetric $SU(5)$

Supersymmetry, i.e. symmetry between bosons and fermions guarantees the cancellation of quadratic divergences for the Higgs mass and thus can make M_W insensitive to M_X. That is, we do not know why M_W/M_X is small, but it is not a problem, since it will stay small in perturbation theory as long as the scale of supersymmetry breaking is small $\Lambda_{SS} \simeq M_W$. The point is that the Higgs mass term is invariant under the internal symmetries and thus is normally not protected from high scales as manifested by quadratic divergences. The fermion masses, on the other hand, are protected by chiral symmetry and thus insensitive to large scales as manifested by 'small' logarithmic divergences. In supersymmetry scalars and fermions are not distinguishable and thus Higgs mass is under control too.

Then for every particle of the standard model there is a supersymmetric partner of the opposite statistics

FERMIONS (quarks, leptons) $s = 1/2$	$\Longleftrightarrow$	SFERMIONS (squarks, sleptons) $s = 0$
GAUGE BOSONS ($W^{\pm}$, Z, γ, gluons) $s = 1$	$\Longleftrightarrow$	GAUGINOS (Wino, Zino, photino, gluinos) $s = 1/2$
HIGGS SCALAR $s = 0$	$\Longleftrightarrow$	HIGGSINO $s = 1/2$

It is easy to see that the formulas for the running of the gauge couplings will be affected by the presence of the new particles. From (90) we get

$$b_G^{SS} = \left(\frac{11}{3} - \frac{2}{3}\right) T_{GB} - \left(\frac{2}{3} + \frac{1}{3}\right) T_F - \left(\frac{1}{3} + \frac{2}{3}\right) T_H \tag{95}$$

or

$$b_G = 3T_{GB} - T_F - T_H \tag{96}$$

where the added contributions in (95) are due to the superpartners.

From (96) we get for the individual gauge couplings

$$\begin{aligned} b_3^{SS} &= 9 - 2n_g \\ b_2^{SS} &= 6 - 2n_g - \frac{1}{2}n_H \\ b_1^{SS} &= -2n_g - \frac{3}{10}n_H \end{aligned} \tag{97}$$

where n_H is again the number of Higgs doublets.

In exactly the same way as before, assuming the unification of couplings at M_X, we find

$$\frac{1}{\alpha_2(M_W)} - \frac{1}{\alpha_3(M_W)} = \frac{6+n_H}{4\pi} \ln \frac{M_X}{M_W}$$

$$\sin^2\theta_W(M_W) = \frac{3}{8} - \frac{30-n_H}{16\pi}\alpha_{em}(M_W) \ln \frac{M_X}{M_W} \tag{98}$$

In the minimal supersymmetric standard model (MSSM) $n_H = 2$, and we find

$$M_X \simeq 10^{16} GeV \tag{99}$$

and

$$\sin^2\theta_W(M_W) = \frac{1}{5} + \frac{7}{15}\frac{\alpha_{em}(M_W)}{\alpha_3(M_W)} \simeq 0.23 \tag{100}$$

MSSM agrees perfectly well with the experiment and with the above value for M_X we predict the proton lifetime

$$\tau_p \simeq 10^{36} yr \tag{101}$$

which is above the experimental bound

$$(\tau_p)_{\rm exp} \geq 6 \cdot 10^{33} yr \tag{102}$$

Now, if we are to take supersymmetry seriously, all the way up to the scale M_X, we expect of course new gauginos $\tilde{X}$, $\tilde{Y}$, associated with the superheavy bosons X and Y of $SU(5)$; and also heavy Higgsinos $\tilde{h}_\alpha$ from **5** of $SU(5)$. The exchange of the heavy Higgsinos leads to proton decay, suppressed only linearly by the GUT scale. More precisely, the exchange of heavy Higgsinos gives the effective operator of the type

$$\frac{1}{M_X} QL\tilde{Q}\tilde{Q} \tag{103}$$

where Q and L stand for quarks and leptons and $\tilde{Q}$ stands for squarks. In turn the squarks are changed into quarks through the exchange of gauginos and one obtains an operator of the form $QQQL$ of the proton decay. While it depends on the Yukawa sector and the sfermion masses and mixings, and thus not easy to predict precisely, proton lifetime is typically very close (or below) the experimental limit. Supersymmetric grand unification, with sfermion and gaugino masses in the TeV region, cries for a new generation of proton decay experiments.

3.5 SU(5) and Neutrino Mass

As we saw above, in the minimal SU(5) neutrinos are massless just as in the SM. Here I discuss a simple extension of the minimal theory which can account for non vanishing neutrino masses. The readers should be warned that this section is much less pedagogical than the previous ones, and a newcomer to the field can skip it in the first reading. What lies ahead may be on the other hand of a particular interest to more learned readers, especially those interested in the physics of neutrino masses and mixings. In the next section on left-right symmetry we will in any case offer a more pedagogical discussion of neutrino mass and the seesaw mechanism.

The seesaw mechanism [23] has been recognized as the most natural scenario for understanding the smallness of neutrino mass. I will discuss it more carefully and more pedagogically in the next section devoted to left-right symmetry and again in the last section devoted to SO(10) grand unified theory. Suffice it to say here that it implies the existence of heavy particles, which after being integrated out, lead to the gauge invariant operator [24]

$$\mathcal{L}_{eff} = y_{eff} \frac{LLHH}{M} , \tag{104}$$

with $M \gg M_W$ usually assumed. Its form is a reflection of the above mentioned fact that neutrino mass term is a SU(2) triplet, and so it is compensated by the SU(2) triplet made out of Higgs doublet squared. Once the Higgs gets a non vanishing vev, neutrino gets a mass small mass

$$m_\nu = y_{eff} \frac{v^2}{M} , \tag{105}$$

where v is the vacuum expectation value of the Higgs doublet H.

As shown in [25], there are only three different types of heavy particles that can induce (104):

I) SM fermionic singlets, coupled to leptons through Dirac Yukawa couplings and usually called right-handed neutrinos (type I seesaw) [23];

II) SU(2) bosonic triplet ($Y = 2$) coupled to leptons through Majorana type couplings (type II seesaw) [26];

III) SU(2) fermionic triplet ($Y = 0$) coupled to leptons through Dirac Yukawas, just like the singlet ones in I) (called type III seesaw) [27].

Exercise 7 *It is nice exercise to show that these heavy states do lead to the effective operator above, once integrated out. Take for example, the type I case, i.e. add the SM model singlet right handed neutrinos* ν_R *and write the usual Dirac Yukawa couplings* Y_D *and the gauge and Lorentz invariant mass terms* $M_R \nu_R^T C \nu_R$. *Show that the effective operator (104) is reproduced with* $y_{eff}/M = Y_D^2/M_R$.

Whatever one chooses, one needs a predictive theory above the SM in order to shed some light on neutrino masses; otherwise, one can as well stick to the effective operator in (104). The best bet for such a theory is grand unification since it can predict new mass scale(s). It turns out that both type I and type II seesaw find their natural role in SO(10) theory due to the automatically present left-right symmetry [28–31]. Although SO(10) is sufficient by itself to determine all the parameters in the I) and II) cases, and even the 1-3 mixing angle [32], the check of the seesaw is only indirect: one can at best relate neutrino properties to proton decay. The main point is that both right-handed neutrinos and the SU(2) scalar triplet are predicted to be very heavy, close to the GUT scale. This is discussed in the last section. In SU(5) the Type II seesaw was studied at great length in [33]. This amounts to the 15_H in addition to the minimal theory and it turns out that some states end up being at intermediate scale. The theory is quite simple and economical, and the most interesting prediction is a fairly low unification scale which makes proton decay likely to be discovered in the future. Still, the seesaw mechanism is not directly testable at collider physics in near future. This seems to be a generic feature of both Type I and II seesaw mechanisms.

What about the type III seesaw? It is clearly custom fit for the SU(5) theory, as suggested recently [34], since it only requires adding the adjoint fermions 24_F to the existing minimal model with three generations of quarks and leptons, and 24_H and 5_H Higgs fields. This automatically leads to the hybrid scenario of both type I and type III seesaw, since 24_F has also a SM singlet fermion, i.e. the right-handed neutrino. This should be clear to the alert student. After all, the 24_F is completely analogous to the 24_F or even better the adjoint gauge boson representation, which we studied at length. The fermionic triplet simply corresponds to the SU(2) gauge boson triplet, whereas the singlet corresponds to the U(1) gauge boson. This singlet can be interpreted as a right handed neutrino, for it is a SM neutral particle with Yukawa couplings to the light neutrinos. The triplet fermion on the other hand has the quantum numbers of the winos, the supersymmetric partners of the SU(2) charged and neutral gauge bosons.

The main prediction of this theory is the lightness of the fermionic triplet. For a conventional value of $M_{GUT} \approx 10^{16}$ GeV, the unification constraints strongly suggest its mass below TeV, relevant for the future colliders such as LHC. The triplet fermion decay predominantly into W (or Z) and leptons, with lifetimes shorter that about 10^{-12} sec.

Equally important, the decays of the triplet are dictated by the same Yukawa couplings that lead to neutrino masses and thus one has an example of predicted low-energy seesaw directly testable at colliders and likely already at LHC.

Unification Constraints and the Mass Scales of the Theory

The minimal implementation of the type III seesaw in non-supersymmetric SU(5) requires a fermionic adjoint 24_F in addition to the usual field content 24_H, 5_H and three generations of fermionic 10_F and $\bar{5}_F$. The consistency of the charged fermion masses requires higher dimensional operators in the usual Yukawa sector [35]. One must add the new Yukawa interactions

$$\begin{aligned}\mathcal{L}_{Y\nu} = \; & y_0^i \bar{5}_F^i 24_F 5_H \\ & + \frac{1}{\Lambda} \bar{5}_F^i \left[y_1^i 24_F 24_H + y_2^i 24_H 24_F + y_3^i Tr\left(24_F 24_H\right)\right] 5_H + h.c. \,.\end{aligned} \tag{106}$$

After the SU(5) breaking one obtains the following physical relevant Yukawa interactions for neutrino with the triplet $T^F \equiv \boldsymbol{T}^F \cdot \boldsymbol{\tau}$ (type III) and singlet S^F (type I) fermions:

$$\mathcal{L}_{Y\nu} = L_i \left(y_T^i T^F + y_S^i S^F \right) H + h.c. \,, \tag{107}$$

where y_T^i, y_S^i are two different linear combinations of y_0^i and $y_a^i v_{GUT}/\Lambda$ ($a = 1,2,3$), L_i are the lepton doublets and H is the Higgs doublet. It is clear from the above formula that besides the new appearance of the triplet fermion, the singlet fermion in 24_F acts precisely as the right-handed neutrino; it should not come out as a surprise, as it has the right SM quantum numbers.

One important prediction emerges immediately: only two light neutrinos get mass, while the third one remains massless. This is seen readily. First, the Yukawas here are vectors, and for example the vector coupling corresponding to the triplet can be rotated in the say 3rd direction. Thus only one light neutrino effectively coupled to the triplet, i.e. only one neutrino gets the mass through this coupling. Obviously, the same could have been said about the singlet, and so we have as argued only two massive light neutrinos. This is of course independent of the nature of the heavy states, and thus the number of light massive neutrinos is in direct proportion to the number of heavy fermions, be they singlets or triplets.

The mass of the fermionic triplet is found by performing the renormalization group analysis as before. From [34] one has

$$\exp\left[30\pi\left(\alpha_1^{-1} - \alpha_2^{-1}\right)(M_Z)\right] = \tag{108}$$

$$\left(\frac{M_{GUT}}{M_Z}\right)^{84} \left(\frac{\left(m_3^F\right)^4 m_3^B}{M_Z^5}\right)^5 \left(\frac{M_{GUT}}{m_{(3,2)}^F}\right)^{20} \left(\frac{M_{GUT}}{m_T}\right) \,,$$

$$\exp\left[20\pi\left(\alpha_1^{-1} - \alpha_3^{-1}\right)(M_Z)\right] = \tag{109}$$

$$\left(\frac{M_{GUT}}{M_Z}\right)^{86} \left(\frac{\left(m_8^F\right)^4 m_8^B}{M_Z^5}\right)^5 \left(\frac{M_{GUT}}{m_{(3,2)}^F}\right)^{20} \left(\frac{M_{GUT}}{m_T}\right)^{-1} \,,$$

where $m_3^{F,B}$, $m_8^{F,B}$, $m_{(3,2)}^F$ and m_T are the masses of weak triplets, color octets, (only fermionic) leptoquarks and (only bosonic) color triplets respectively.

We discussed at length the well known problem in the standard model of the low meeting scale of α_1 and α_2. It is clear that the SU(2) triplet fermions are ideal from this point of view since they slow down the running of α_2, while leaving α_1 intact (other particles have non vanishing hypercharge and thus make α_1 grow faster as to meet α_2 even before). They should clearly be as light as possible while the color triplet as heavy as possible. In order to illustrate the point, take $m_3^F = m_3^B = M_Z$ and $m_T = M_{GUT}$. This gives ($\alpha_1^{-1}(M_Z) = 59$, $\alpha_2^{-1}(M_Z) = 29.57$, $\alpha_3^{-1}(M_Z) = 8.55$) $M_{GUT} \approx 10^{15.5}$ GeV. Increasing the triplet masses $m_3^{F,B}$ reduces M_{GUT} dangerously, making proton decay too fast.

Finally, one can ask, where must the octets be. Since the triplets slowed down the running of α_2, the meting point of α_2 and α_3 would become too large, unless α_3 gets slowed down too. Thus the octets must lie much below M_{GUT}, but since they contribute to the running more than the triplets, they should be also much above the weak scale, i.e. we expect them at the intermediate scale.

For a more detailed discussion of unification constraints and especially the phenomenology of the triplet relevant for LHC see [36].

4 Left-right Symmetry and Neutrino Mass

This section is devoted to the left-right symmetric extension of the standard model and the issue of the origin of the breaking of parity. I include it in these lectures since L-R symmetry is a finite gauge symmetry in SO(10) in a form of charge conjugation. I recommend its reading mostly to the students interested in neutrino masses and mixings for it facilitates understanding of these issues in SO(10), and also to those who believe that the crippled, purely left form of the standard model must be modified at large energies. The write-up here is rather simple and pedagogical, without too many technicalities.

4.1 Parity

Parity is the fundamental symmetry between left and right, the first symmetry a child notices, and its breaking, I believe, should be understood, or in other words. In the standard model P is broken explicitly and clearly, in order to break P spontaneously we must enlarge the gauge group. The minimal model is based on the gauge group [28–31].

$$G_{LR} = SU(2)_L \times SU(2)_R \times U(1)_{Y'}$$

with the quarks and leptons completely symmetric under $L \leftrightarrow R$

$$\begin{aligned} Q_L = \begin{pmatrix} u \\ d \end{pmatrix}_L &\xleftrightarrow{P} Q_R = \begin{pmatrix} u \\ d \end{pmatrix}_R \\ \ell_L = \begin{pmatrix} \nu \\ e \end{pmatrix}_L &\xleftrightarrow{P} \ell_R = \begin{pmatrix} \nu \\ e \end{pmatrix}_R \end{aligned} \tag{110}$$

Notice that the requirement of left-right symmetry leads to the existence of the right-handed neutrino and now the neutrino mass becomes a dynamical issue, related to the pattern of symmetry breaking. In the Standard Model, where ν_R is absent, $m_\nu = 0$; here instead we shall need to explain why neutrinos are so much lighter than the corresponding charged leptons.

In this theory, the formula for the electromagnetic charge becomes

$$Q_{em} = I_{3L} + I_{3R} + \frac{B-L}{2} \tag{111}$$

This is in sharp contrast with the Standard Model, where the hypercharge Y was completely devoid of any physical meaning. So L-R symmetry is deeply connected with B-L symmetry; the existence of right-handed neutrinos implied by L-R symmetry is necessary in order to cancel anomalies when gauging B-L. Namely, the B-L symmetry is a global anomaly free symmetry of the SM, but without ν_R the gauged version would have $(B-L)^3$ anomaly.

Our primary task is to break L-R symmetry, i.e. to account for the fact that $M_{W_R} \gg M_{W_L}$, W_R and W_L denoting right-handed and left-handed gauge bosons respectively. In order to do so we need a set of left-handed and right-handed Higgs scalars whose quantum numbers we will specify later. Imagine for the moment two scalars φ_L and φ_R with

$$\varphi_L \overset{P}{\longleftrightarrow} \varphi_R \tag{112}$$

Assume no terms linear in the fields (since φ_L and φ_R should carry quantum numbers under $SU(2)_L$ and $SU(2)_R$) we can write down the left-right symmetric potential

$$V = -\frac{\mu^2}{2}(\varphi_L^2 + \varphi_R^2) + \frac{\lambda}{4}(\varphi_L^4 + \varphi_R^4) + \frac{\lambda'}{2}\varphi_L^2\varphi_R^2 \tag{113}$$

where $\lambda > 0$ in order for V to be bounded from below, and we choose $\mu^2 > 0$ in order to achieve symmetry breaking in the usual manner. We rewrite the potential as

$$V = -\frac{\mu^2}{2}(\varphi_L^2 + \varphi_R^2) + \frac{\lambda}{4}(\varphi_L^2 + \varphi_R^2)^2 + \frac{\lambda' - \lambda}{2}\varphi_L^2\varphi_R^2 \tag{114}$$

which tells us that the pattern of symmetry breaking depends crucially on the sign of $\lambda' - \lambda$, since the first two terms do not depend on the direction of symmetry breaking (of course $\mu^2 > 0$ guarantees that $<\varphi_L> = <\varphi_R> = 0$ is a maximum and not a minimum of the potential).

Exercise 8 *Show that if*

1. $\lambda' - \lambda > 0$, *in order to minimize* V *we have either* $<\varphi_L> = 0$, $<\varphi_R> \neq 0$, *or vice versa. Due to the symmetry of* V *both solutions are equally probable.*
2. $\lambda' - \lambda < 0$, *we need* $<\varphi_L> \neq 0 \neq <\varphi_R>$ *and L-R symmetry implies* $<\varphi_L> = <\varphi_R>$.

Obviously we choose 1., which implies that P is broken in nature [30, 31].

4.2 Left-Right Symmetry and Massive Neutrinos

What fields should we choose for the role of φ_L and φ_R? From the neutrino mass point of view, the ideal candidates should be triplets, i.e.

$$\Delta_L(\bar{3}_L, 1_R, 2) \quad ; \quad \Delta_R(\bar{1}_L, 3_R, 2) \tag{115}$$

where the quantum numbers denote $SU(2)_L$, $SU(2)_R$ and $B-L$ transformation properties. Simply speaking, Δ_L and Δ_R are $SU(2)_L$ and $SU(2)_R$ triplets, respectively, with $B-L$ numbers equal to two.

Writing $\Delta_{L,R} = \Delta^i_{L,R}\tau_i/2$ (τ_i being the Pauli matrices) as is usual for the adjoint representations, we find Yukawa couplings

$$\mathcal{L}_\Delta = h_\Delta(\ell_L^T\, C\, i\tau_2\, \Delta_L\, \ell_L + L \to R) + h.c. \tag{116}$$

To check the invariance of (116) under the Lorentz group and the gauge symmetry $SU(2)_L \times SU(2)_R \times U(1)_{B-L}$, recall

- that $\psi_L^T C \psi_L$ is a Lorentz invariant quantity for a chiral Weyl spinor ψ_L (and similarly for ψ_R).
- under the gauge symmetry $SU(2)_L$

$$\ell_L \longrightarrow \mathcal{U}_L \ell_L \quad , \quad \Delta_L \longrightarrow \mathcal{U}_L \Delta_L \mathcal{U}_L^\dagger$$
$$\mathcal{U}_L^T(i\tau_2) = (i\tau_2)\mathcal{U}_L^\dagger \tag{117}$$

  and similarly for $SU(2)_R$
- the B-L number of the $\Delta_{L,R}$ fields is two.

This proves the invariance of (116) under all the relevant symmetries. Now, from their definition, the fields $\Delta_{L,R}$ have the following decomposition under the charge eigenstates

$$\Delta_{L,R} = \begin{bmatrix} \Delta^+/\sqrt{2} & \Delta^{++} \\ \Delta^0 & -\Delta^+/\sqrt{2} \end{bmatrix}_{L,R} \tag{118}$$

where we use the fact that $Tr\Delta_{L,R} = 0$ and the charge is computed from $Q = I_{3L} + I_{3R} + (B-L)/2$.

Notice an interesting consequence of doubly charged physical Higgs scalars in this theory. From the general analysis of the spontaneous L-R symmetry breaking, we know that for a range of parameters of the potential the minimum of the theory can be chosen as

$$\langle \Delta_L \rangle = 0 \quad , \quad \langle \Delta_R \rangle = \begin{bmatrix} 0 & 0 \\ v_R & 0 \end{bmatrix} \tag{119}$$

From (116), we the obtain the mass for the right-handed neutrino ν_R

$$\mathcal{L}_m = h_\Delta\, v_R\, (\nu_R^T\, C\, \nu_R + \nu_R^\dagger\, C^\dagger\, \nu_R^*) \tag{120}$$

Thus the right-handed neutrino gets a large mass $M_R = h_\Delta v_R$, which corresponds to the scale of breaking of parity. At the same time, the original gauge symmetry is broken down to the Standard Model one

$$SU(2)_L \times SU(2)_R \times U(1)_{B-L} \overset{<\Delta_R>}{\longrightarrow} SU(2)_L \times U(1)_Y \tag{121}$$

This can be checked by computing the gauge boson mass matrix. By defining the right-handed charged gauge boson

$$W_R^{\pm} = \frac{A_R^1 \mp i A_R^2}{\sqrt{2}} \tag{122}$$

we get

$$M_{W_R}^2 = g^2 \, v_R^2 \tag{123}$$

$$M_{Z_R}^2 = (g^2 + 4 g_{B-L}^2) \, v_R^2 \tag{124}$$

where

$$Z_R = \frac{2 g_{B-L} A_R^3 + g A_{B-L}}{\sqrt{g^2 + 4 g_{B-L}^2}} \tag{125}$$

is the new massive neutral gauge field.

Thus the scale of parity breaking is related to the mass of the right-handed charged gauge bosons $W_R^{\pm}$. The predominant V-A nature of the weak interactions puts a lower limit on M_{W_R}

$$M_{W_R} > 1 TeV \tag{126}$$

To complete the theory, one needs a Higgs bidoublet which contains the SM Higgs, so that one can give masses to quarks and leptons. In the process we get the Dirac neutrino mass between ν_L and ν_R and in turn we end up with the see-saw mechanism for light neutrino masses. This is conventionally called the Type I see-saw.

Type I see-saw

From the Dirac Yukawas

$$\mathcal{L} = h_{\Phi} \overline{\ell_L} \, \Phi \, \ell_R + h.c. \tag{127}$$

after the symmetry breaking the neutrino Dirac mass term is $m_D = h_{\Phi} \langle \Phi \rangle$. The neutrino mass terms become

$$m_D \overline{\ell_L} \ell_R + M_R \ell_R^T C \ell_R + h.c. \tag{128}$$

and the neutrino mass matrix looks like

$$\begin{pmatrix} 0 & m_D \\ m_D & M_R \end{pmatrix} \tag{129}$$

therefore its eigenvalues are

$$m_{\pm} = \frac{M_R}{2} \pm \frac{M_R}{2}\sqrt{1 + \frac{m_D^2}{M_R^2}} \tag{130}$$

Since

$$M_R \gg m_D \tag{131}$$

we have from (130) a heavy state with

$$m_{\nu_+} \sim M_R \tag{132}$$

consisting essentially of ν_R with little admixture of ν_L, and a light state

$$m_{\nu_-} \sim \frac{m_D^2}{M_R} \tag{133}$$

made up almost entirely of ν_L. This is the famous *see-saw* mechanism: a right-handed neutrino, neutral under the SM group and with a large Majorana mass, forces the left-handed neutrino to be very light in comparison with the charged fermions.

Notice that upon integrating out the heavy field ν_R, one obtains the light neutrino mass matrix

$$M_\nu = m_D^T \frac{1}{M_R} m_D \tag{134}$$

which is effective operator of Weinberg in 104.

The important point here is that the mass of ν_R is determined by the scale of parity breaking and the smallness of the neutrino mass is a reflection of the predominant V-A structure of the weak interaction and provides a probe of parity restoration at high energies $E > M_{W_R}$.

Type II see-saw

The gauge symmetry of the Left-Right model allows also for the following term in the potential that we have ignored before for simplicity

$$\Delta V = \alpha \Delta_L^\dagger \Phi \Delta_R \Phi^\dagger \tag{135}$$

which implies that $\langle \Delta_L \rangle$ cannot vanish.

Exercise 9 *Show that*

$$\langle \Delta_L \rangle \simeq \alpha \frac{M_W^2 \langle \Delta_R \rangle}{M_{\Delta_L}} \simeq \alpha \frac{M_W^2}{M_R} \tag{136}$$

The tiny vev of Δ_L has only impact on the neutrino mass. From (116) neutrino gets a small mass

$$m_\nu = h_\Delta \langle \Delta_L \rangle \tag{137}$$

which is of the same order as the Type I see-saw contribution. This direct mass is referred to as Type II see-saw.

The predictions for neutrino mass depend crucially on M_{W_R}, but the L-R symmetric model by itself cannot give us its value. This is cured in SO(10) grand unified theory, where we will see that this scale is very large which fits perfectly with observed neutrino masses. Even more important, in SO(10) this symmetry introduced here ad-hoc, is an automatic finite gauge transformation.

5 SO(10)

The minimal gauge group that unifies the gauge interactions of the standard model was seen in the previous subsection to be based on SU(5) and studied at length. It is tailor fit for massless neutrinos just as the SM, for in the minimal version of the theory neutrinos get neither Dirac nor Majorana mass terms. Furthermore, the ordinary, non supersymmetric theory fails to unify gauge couplings. We found that the simple extension with the adjoint fermion representation provides a minimal and remarkably predictive theory with light fermionic triplet expected at LHC and whose decay rates probe the Dirac Yukawa couplings of neutrinos. We have a theory that works and furthermore gives serious hope for an old dream of verifying seesaw mechanism at colliders. So why should one ever wish to go beyond SU(5)? We can think of at least two reasons. First, if one is to worry about the Higgs mass naturalness, one may wish to include supersymmetry. While SU(5) with the low energy supersymmetry has a rather appealing feature of providing automatically (as predicted many years ago) a gauge coupling unification, it is not an interesting theory of fermion masses and mixings. First of all, it offers no explanation for the smallness of R-parity violation in nature, and at the same time it requires a certain amount of arbitrary and unpredicted R-parity violation in order to provide neutrino masses. One can also include the type II seesaw into the theory through the 15_H supermultiplet, and even attribute to it a mediation of supersymmetry breaking, but one ends up without any direct low energy probes or interesting quark-lepton mass and mixings relations. This is where SO(10) fits ideally, for it also unifies matter besides the interactions. It works nicely without supersymmetry too, for it provides a natural unification of gauge couplings through the intermediate scale of L-R symmetry breaking.

We discussed the group theory of SO(2N) in Sect. 1. The one important representation of SO(10) is a 16-dimensional spinor, which can be decomposed under SU(5) as $16 = 10 + \bar{5} + 1$. It unifies a family of fermions with an addition of a right handed neutrino per family. This minimal grand unified theory that unifies matter on top of interactions suggests naturally small neutrino masses through the seesaw mechanism. Furthermore, it relates neutrino masses and mixings to the ones of charged fermions, and is predictive in its minimal version. In this section I discuss some salient features in this theory while focusing on its minimal realizations. The crucial representation is a self-dual five index anti-symmetric one responsible for right-handed neutrino masses and is a must, whether being elementary of composed at the loop level or through the higher dimensional operators. A number of different minimal realizations of SO(10) depends on this construction, and what follows summarizes a few of them. The material of Sect. 1 should be sufficient to follow the rest.

There are a number of features that make SO(10) special:

1. a family of fermions is unified in a 16-dimensional spinorial representation; this in turn predicts the existence of right-handed neutrinos
2. $L-R$ symmetry is a finite gauge transformation in the form of charge conjugation. This is a consequence of both left-handed fermions f_L and its charged conjugated counterparts $(f^c)_L \equiv C\overline{f}_R^T$ residing in the same representation 16_F.
3. in the supersymmetric version, matter parity $M=(-1)^{3(B-L)}$, equivalent to the R-parity $R=M(-1)^{2S}$, is a gauge transformation [42], a part of the center Z_4 of SO(10). It simply reads $16 \to -16$, $10 \to 10$. Its fate depends then on the pattern of symmetry breaking (or the choice of Higgs fields); it turns out that in the renormalizable version of the theory R-parity remains exact at all energies [43,44]. The lightest supersymmetric partner (LSP) is then stable and is a natural candidate for the dark matter of the universe.
4. its other maximal subgroup, besides $SU(5)\times U(1)$, is $SO(4)\times SO(6) = SU(2)_L\times SU(2)_R\times SU(4)_c$ symmetry of Pati and Salam. It explains immediately the somewhat mysterious relations $m_d = m_e$ (or $m_d = 1/3 m_e$) of SU(5).
5. the unification of gauge couplings can be achieved with or without supersymmetry.
6. the minimal renormalizable version (with no higher dimensional $1/M_{Pl}$ terms) offers a simple and deep connection between $b-\tau$ unification and a large atmospheric mixing angle in the context of the type II see-saw [45].

In order to understand some of these results, and in order to address the issue of construction of the theory, we turn now to the Yukawa sector.

5.1 Yukawa Sector

Fermions belong to the spinor representation 16_F[1] [20]. From

$$16\times 16 = 10 + 120 + 126 \tag{138}$$

[1] The group theory material of the first section is of particular use for the discussion of spinors and the rest of this section.

the most general Yukawa sector in general contains 10_H, 120_H and $\overline{126}_H$, respectively the fundamental vector representation, the three-index antisymmetric representation and the five-index antisymmetric and anti-self-dual representation. This can be seen by analogy with the Yukawa couplings of $SO(6)$ (see Sect. 2),

$$\begin{aligned}\mathcal{L}_y = \;& y_{10}\Psi^T B\Gamma_i\Psi\Phi_i + y_{120}\Psi^T B\Gamma_i\Gamma_j\Gamma_k\Psi\Phi_{[ijk]} \\ & + y_{126}\Psi^T B\Gamma_i\Gamma_j\Gamma_k\Gamma_l\Gamma_m\Psi\Phi^-_{[ijklm]}\end{aligned} \tag{139}$$

$\overline{126}_H$ is necessarily complex, supersymmetric or not; 10_H and $\overline{126}_H$ Yukawa matrices are symmetric in generation space, while the 120_H one is antisymmetric.

Understanding fermion masses is easier in the Pati-Salam language of one of the two maximal subgroups of SO(10), $G_{PS} = SU(4)_c \times SU(2)_L \times SU(2)_R$ (the other being $SU(5)\times U(1)$). Let us decompose the relevant representations under G_{PS}

$$\begin{aligned}\mathbf{16} &= (4,2,1) + (\bar{4},1,2) \\ \mathbf{10} &= (1,2,2) + (6,1,1) \\ \mathbf{120} &= (1,2,2) + (6,3,1) + (6,1,3) + (15,2,2) + (10,1,1) + (\overline{10},1,1) \\ \overline{\mathbf{126}} &= (\overline{10},3,1) + (10,1,3) + (15,2,2) + (6,1,1)\end{aligned} \tag{140}$$

Exercise 10 *Try to arrive at the above decomposition using the material in Sect. 2.*

Clearly, the see-saw mechanism, whether type I or II, requires $\overline{126}$: it contains both $(10,1,3)$ whose vev gives a mass to ν_R (type I), and $(\overline{10},3,1)$, which contains a color singlet, $B-L=2$ field Δ_L, that can give directly a small mass to ν_L (type II). A reader familiar with the SU(5) language sees this immediately from the decomposition under this group

$$\overline{\mathbf{126}} = 1 + 5 + 15 + \overline{45} + 50 \tag{141}$$

The 1 of SU(5) belongs to the $(10,1,3)$ of G_{PS} and gives a mass for ν_R, while 15 corresponds to the $(\overline{10},3,1)$ and gives the direct mass to ν_L.

Of course, $\overline{126}_H$ can be a fundamental field, or a composite of two $\overline{16}_H$ fields, or can even be induced as a two-loop effective representation built out of a 10_H and two gauge 45-dim representations. In what follows I shall discuss carefully all three possibilities.

Normally the light Higgs is chosen to be the smallest one, 10_H. Since $\langle 10_H\rangle = \langle (1,2,2)\rangle_{PS}$ is a $SU(4)_c$ singlet, $m_d = m_e$ follows immediately, independently of the number of 10_H you wish to have. Thus we must add either 120_H or $\overline{126}_H$ or both in order to correct the bad mass relations. Both of these fields contain $(15,2,2)_{PS}$, and its vev gives the relation $m_e = -3m_d$.

As $\overline{126}_H$ is needed anyway for the see-saw, it is natural to take this first. The crucial point here is that in general $(1,2,2)$ and $(15,2,2)$ mix through

$\langle(10,1,3)\rangle$ [46] and thus the light Higgs is a mixture if the two. In other words, $\langle(15,2,2)\rangle$ in $\overline{126}_H$ is in general non-vanishing[2]. It is rather appealing that 10_H and $\overline{126}_H$ may be sufficient for all the fermion masses, with only two sets of symmetric Yukawa coupling matrices.

5.2 An Instructive Failure

Before proceeding, let me emphasize the crucial point of the necessity of 120_H or $\overline{126}_H$ in the charged fermion sector on an instructive failure: a simple and beautiful model by Witten [47]. The model is non-supersymmetric and the SUSY lovers may place the blame for the failure here. It uses $\langle 16_H\rangle$ in order to break $B-L$, and the "light" Higgs is 10_H. Witten noticed an ingenious and simple way of generating an effective mass for the right-handed neutrino, through a two-loop effect which gives

$$M_{\nu_R} \simeq y_{up}\left(\frac{\alpha}{\pi}\right)^2 M_{GUT} \tag{142}$$

where one takes all the large mass scales, together with $\langle 16_H\rangle$, of the order M_{GUT}. Since $\langle 10_H\rangle = \langle(1,2,2)_{PS}\rangle$ preserves quark-lepton symmetry, it is easy to see that

$$\begin{aligned} M_\nu &\propto M_u \\ M_e &= M_d \\ M_u &\propto M_d \end{aligned} \tag{143}$$

so that $V_{\text{lepton}} = V_{\text{quark}} = 1$. The model fails badly.

The original motivation of Witten was a desire to know the scale of M_{ν_R} and increase M_ν, at that time neutrino masses were expected to be larger. But the real achievement of this simple, elegant, minimal SO(10) theory is the predictivity of the structure of M_{ν_R} and thus M_ν. It is an example of a good, albeit wrong theory: it fails because it predicts.

What is the moral behind the failure? Not easy to answer. The main problem, in my opinion, was to ignore the fact that with only 10_H already charged fermion masses fail. One needs to enlarge the Higgs sector, by adding for example a 120_H); the theory still leads to interesting predictions while possible completely realistic.

5.3 Non-supersymmetric $SO(10)$

In the last two decades, and especially after its success with gauge coupling unification, grand unification by an large got tied up with low energy supersymmetry. This is certainly well motivated, since supersymmetry is the only

[2] In supersymmetry this is not automatic, but depends on the Higgs superfields needed to break SO(10) at M_{GUT}.

mechanism in field theory which controls the gauge hierarchy. In $SO(10)$, gauge coupling unification needs no supersymmetry whatsoever. It only says that there must be intermediate scales [48], such as Pati-Salam $SU(4)_c \times SU(2)_L \times SU(2)_R$ or Left-Right $SU(3)_c \times SU(2)_L \times SU(2)_R \times U(1)_{B-L}$ symmetry, between M_W and M_{GUT}. An oasis or two in the desert is always welcome.

Thus if we accept the fine-tuning, as we seem to be forced in the case of the cosmological constant, we can as well study the ordinary, non-supersymmetric version of the theory. In this context the idea of the cosmic attractors [49] as the solution to the gauge hierarchy becomes extremely appealing. It needs no supersymmetry whatsoever, and enhances the motivation for ordinary grand unified theories. In what follows I discuss some essential features of a possible minimal such theory with 126_H as a necessary ingredient for see-saw.

Let us start by analyzing the case with an extra 10_H field [61]. The most general Yukawa interaction is

$$\mathcal{L}_Y = 16_F \left(10_H Y_{10} + \overline{126}_H Y_{126}\right) \mathbf{16_F} + h.c. \,. \tag{144}$$

where Y_{10} and Y_{126} are symmetric matrices in the generation space. With this one obtains relations for the Dirac fermion masses

$$\begin{aligned} M_D &= M_1 + M_0, & M_U &= c_1 M_1 + c_0 M_0, \\ M_E &= -3M_1 + M_0, & M_{\nu_D} &= -3c_1 M_1 + c_0 M_0, \end{aligned} \tag{145}$$

where we have defined

$$M_1 = \langle 2,2,15\rangle^d_{126}\, Y_{126}\ , \quad M_0 = \langle 2,2,1\rangle^d_{10}\, Y_{10}\ , \tag{146}$$

and

$$c_0 = \frac{\langle 2,2,1\rangle^u_{10}}{\langle 2,2,1\rangle^d_{10}}\ ,\ c_1 = \frac{\langle 2,2,15\rangle^u_{126}}{\langle 2,2,15\rangle^d_{126}}\ . \tag{147}$$

In the physically sensible approximation $\theta_q = V_{cb} = 0$, these relations imply

$$c_0 = \frac{m_c(m_\tau - m_b) - m_t(m_\mu - m_s)}{m_s m_\tau - m_\mu m_b} \approx \frac{m_t}{m_b}\ , \tag{148}$$

Exercise 11 *Derive this formula.*

Notice that this means that 10_H cannot be real, since in that case one would have $|\langle 2,2,1\rangle^u_{10}| = |\langle 2,2,1\rangle^d_{10}|$, implying m_t/m_b of order one. It is necessary to complexify 10_H, just as in a supersymmetric theory. If taking advantage of this fact one decides to impose a Peccei-Quinn symmetry, thus providing a Dark Matter candidate, the Yukawa sector in non-supersymmetric and supersymmetric models is similar.

In this case, this model has the interesting feature of automatic connection between $b-\tau$ unification and large atmospheric mixing angle in the type II see-saw. From $M_{\nu_L} \propto Y_{126}$, one has $M_{\nu_L} \propto M_D - M_E$. as shown in [45, 50].

This fact has inspired the careful study of the analogous supersymmetric version where $m_\tau \simeq m_b$ at the GUT scale works rather well [55]. In the non-supersymmetric theory, $b - \tau$ unification fails badly, $m_\tau \sim 2m_b$ [56]. The realistic theory will require a Type I seesaw, or an admixture of both possibilities.

Suppose now that we choose instead $\mathbf{120_H}$ [61]. Since Y_{120} is antisymmetric, this means only 3 new complex couplings on top of Y_{126}. On gets in this case

$$\begin{aligned} M_D &= M_1 + M_2 \ , & M_U &= c_1 M_1 + c_2 M_2 \ , \\ M_E &= -3M_1 + c_3 M_2 \ , & M_{\nu_D} &= -3c_1 M_1 + c_4 M_2 \end{aligned} \tag{149}$$

where M_1 and c_1 are defined in (146),(147), and:

$$\begin{aligned} M_2 &= Y_{120}\left(\langle 2,2,1\rangle^d_{120} + \langle 2,2,15\rangle^d_{120}\right) \ , & c_2 &= \frac{\langle 2,2,1\rangle^u_{120} + \langle 2,2,15\rangle^u_{120}}{\langle 2,2,1\rangle^d_{120} + \langle 2,2,15\rangle^d_{120}} \ , \\ c_3 &= \frac{\langle 2,2,1\rangle^d_{120} - 3\langle 2,2,15\rangle^d_{120}}{\langle 2,2,1\rangle^d_{120} + \langle 2,2,15\rangle^d_{120}} \ , & c_4 &= \frac{\langle 2,2,1\rangle^u_{120} - 3\langle 2,2,15\rangle^u_{120}}{\langle 2,2,1\rangle^d_{120} + \langle 2,2,15\rangle^d_{120}} \ . \end{aligned} \tag{150}$$

It is easy to see that again there is a need to complexify the Higgs fields, by arguments similar to the case of 10_H.

In order to obtain algebraic expressions, from which a clearer physical meaning can be extracted, one can restrict the analysis to the second and third generations. Later, numerical studies could include the effects of the first generation as a perturbation. In the basis where M_1 is diagonal, real and non-negative, for the two-generation case one gets:

$$M_1 \propto \begin{pmatrix} \sin^2\theta & 0 \\ 0 & \cos^2\theta \end{pmatrix} \tag{151}$$

and the most general charged fermion matrix can be written as:

$$M_f = \mu_f \begin{pmatrix} \sin^2\theta & i(\sin\theta\cos\theta + \epsilon_f) \\ -i(\sin\theta\cos\theta + \epsilon_f) & \cos^2\theta \end{pmatrix} \ , \tag{152}$$

where $f = D, U, E$ stands for charged fermions and ϵ_f vanishes for negligible second generation masses. In other words $|\epsilon_f| \propto m_2^f / m_3^f$. Furthermore the real parameter μ_f sets the third generation mass scale. By calculating up to leading order in $|\epsilon_f|$, we have to the following interesting predictions [61]:

1. type I and type II seesaw lead to the same structure

$$M_N^I \propto M_N^{II} \propto M_1 \tag{153}$$

so that in the selected basis the neutrino mass matrix is diagonal. We see that the angle θ has to be identified with the leptonic (atmospheric) mixing angle θ_A up to terms of the order of $|\epsilon_E| \approx m_\mu/m_\tau$. For the neutrino masses we obtain from (151)

$$\frac{m_3^2 - m_2^2}{m_3^2 + m_2^2} = \frac{\cos 2\theta_A}{1 - \sin^2 2\theta_A/2} + \mathcal{O}(|\epsilon|) \tag{154}$$

Exercise 12 *Derive this formula.*
This equation points to an intriguing correlation: the degeneracy of neutrino masses is measured by the maximality of the atmospheric mixing angle.

2. the ratio of tau and bottom mass at the GUT scale is given by:

$$\frac{m_\tau}{m_b} = 3 + \mathcal{O}(|\epsilon|) \tag{155}$$

This is not correct in principle, the extrapolation in standard model gives $m_\tau \approx 2m_b$. However, several effects modify this conclusion, such as for example the inclusion of the first generation or the running of Yukawa couplings. We would in any case expect that m_b comes out as small as possible.

3. the quark mixing is found to be:

$$|V_{cb}| = |\cos 2\theta_A\ (\epsilon_D - \epsilon_U)| + \mathcal{O}(|\epsilon^2|) \tag{156}$$

This equation demonstrates the successful coexistence of small and large mixing angles. In order for it to work quantitatively, $|\cos 2\theta_A|$ should be as large as possible, i.e. θ_A should be as far as possible from the maximal value 45°. To make a definite numerical statement, again, the effects from the first generation and the loops have to be included.

5.4 Supersymmetric Case

In supersymmetry 10_H is necessarily complex and the bidoublet $(1, 2, 2)$ in 10_H contains the two Higgs doublets of the MSSM, with the vevs v^u and v^d in general different: $\tan\beta \equiv v^u/v^d \neq 1$ in general. In order to study the physics of SO(10), we need to know what the theory is, i.e. its Higgs content. There are two orthogonal approaches to the issue, as we discuss now.

bf Small representations The idea: take the smallest Higgs fields (least number of fields, not of representations) that can break SO(10) down to the MSSM and give realistic fermion masses and mixings. The following fields are both necessary and sufficient

$$45_H, 16_H + \overline{16}_H, 10_H \tag{157}$$

It all looks simple and easy to deal with, but the superpotential becomes extremely complicated. First, at the renormalizable level it is too simple. The pure Higgs and the Yukawa superpotential at the renormalizable level take the form

$$\begin{aligned} W_H = & m_{45} 45_H^2 + m_{16} 16_H \overline{16}_H + \lambda_1 16_H \Gamma^2 \overline{16}_H 45_H \\ & m_{10} 10_H^2 + \lambda_2 16_H \Gamma 16_F 10_H + \lambda_3 \overline{16}_H \Gamma \overline{16}_H 10_H \end{aligned} \tag{158}$$

$$W_y = y_{10} 16_F \Gamma 16_F 10_H \tag{159}$$

where Γ stands for the Clifford algebra matrices of SO(10), $\Gamma_1 ... \Gamma_{10}$, and the products of Γ's are written in a symbolic notation (both internal and Lorentz charge conjugation are omitted).

Clearly, both W_H and W_y are insufficient. The fermion mass matrices would be completely unrealistic and the vevs $\langle 45_H \rangle, \langle 6_H \rangle, \langle \overline{16}_H \rangle$ would all point in the SU(5) direction. Thus, one adds non-renormalizable operators

$$\begin{aligned} \Delta W_H = & \frac{1}{M_{Pl}} \left[(45_H^2)^2 + 45_H^4 + (16_H \overline{16}_H)^2 + (16_H \Gamma^2 \overline{16}_H)^2 + (16_H \Gamma^4 \overline{16}_H)^2 \right. \\ & + (16_H \Gamma 16_H)^2 + (16_H \Gamma^5 16_H)^2 + \{16_H \to \overline{16}_H\} \\ & \left. + 16_H \Gamma^4 \overline{16}_H 45_H^2 + 16_H \Gamma^3 \overline{16}_H 45_H 10_H + \{16_H \to \overline{16}_H\} \right] \end{aligned} \tag{160}$$

$$\begin{aligned} \Delta W_y = & \frac{1}{M_{Pl}} \left[16_F \Gamma 16_F \, 16_H \Gamma 16_H + \{16_H \to \overline{16}_H\} \right. \\ & \left. 16_F \Gamma^3 16_F 45_H 10_H + 16_F \Gamma^5 16_F \overline{16}_H \Gamma^5 \overline{16}_H \right] \end{aligned} \tag{161}$$

where I take for simplicity all the couplings to be unity; there are simply too many of them. The large number of Yukawa couplings means very little predictivity.

The way out is to add flavor symmetries and to play the texture game and thus reduce the number of couplings. This in a sense goes beyond grand unification and appeals to new physics at M_{Pl} and/or new symmetries.

To me, maybe the least appealing aspect of this approach is the loss of R (matter) parity due to 16_H and $\overline{16}_H$; it must be postulated by hand as much as in the MSSM.

On the positive side, it is an asymptotically free theory and one can work in the perturbative regime all the way up to M_{Pl}. While this sounds nice, I am not sure what it means in practice. It would be crucial if you were able to make high precision determination of M_{GUT} or m_T, the mass of colored triplets responsible for $d = 5$ proton decay. The trouble is that the lack of knowledge of the superpotential couplings is sufficient even in the minimal SU(5) theory to prevent this task; in SO(10) it gets even worse.

Maybe more relevant is the fact that in this scenario $M_R \simeq M_{GUT}^2/M_{Pl} \simeq 10^{13} - 10^{14} GeV$, which fits nicely with the neutrino masses via see-saw. Furthermore, see-saw can be considered "clean", of the pure type I, since the type II effect is suppressed by $1/M_{Pl}$. Most important, the $m_b \simeq m_\tau$ relation from (159) is maintained due to small $1/M_{Pl}$ effects relevant only for the first two generations.

Large representations

The non-renormalizable operators in reality mean invoking new physics beyond grand unification. This may be necessary, but still, one should be more ambitious and try to use the renormalizable theory only. This means large representations necessarily: at least $\overline{126}_H$ is needed in order to give the mass to ν_R (in supersymmetry, one must add 126_H). The consequence is the loss of asymptotic freedom above M_{GUT}, the coupling constants grow large at the scale $\Lambda_F \simeq 10 M_{GUT}$.

Once we accept large representations, we should minimize their number. The minimal theory contains, on top of 10_H, 126_H and $\overline{126}_H$, also 210_H [57–60] with the decomposition

$$\begin{aligned} 210_H = &(1,1,1)_- + (15,1,1)_+ + (15,1,3) + (15,3,1) \\ &+(6,2,2) + (10,2,2) + (\overline{10},2,2) \end{aligned} \tag{162}$$

where the -(+) subscript denotes the properties of the color singlets under charge conjugation.

The Higgs superpotential is remarkably simple

$$\begin{aligned} W_H = &m_{210}(210_H)^2 + m_{126}\overline{126_H}126_H + m_{10}(10_H)^2 + \lambda(210_H)^3 \\ &+\eta 126_H\overline{126}_H 210_H + \alpha 10_H 126_H 210_H + \overline{\alpha} 10_H \overline{126}_H 210_H \end{aligned} \tag{163}$$

and the Yukawa one even simpler

$$W_Y = y_{10} 16_F \Gamma 16_F 10_H + y_{126} 16_F \Gamma^5 16_F \overline{126}_H \tag{164}$$

Remarkably enough, this may be sufficient, without any higher dimensional operators; however, the situation is not completely clear.

There is a small number of parameters: 3 + 6x2 = 15 real Yukawa couplings, and 11 real parameters in the Higgs sector. In this sense the theory can be considered as the minimal supersymmetric GUT in general [60]. As usual, I am not counting the parameters associated with the SUSY breaking terms.

The nicest feature of this program (and the best justification for the use of large representations) is the following. Besides the $\langle(10,1,3)\rangle$ which gives masses to the ν_R's, also the $\langle(15,2,2)\rangle$ in $\overline{126}_H$ gets a vev [46, 58]. Approximately

$$\langle 15,2,2\rangle_{\overline{126}} \simeq \frac{M_{PS}}{M_{GUT}} \langle 1,2,2\rangle \tag{165}$$

with $M_{PS} = \langle 15,2,2\rangle$ being the scale of $SU(4)_c$ symmetry breaking. In SUSY, $M_{PS} \leq M_{GUT}$ and thus one can have correct mass relations for the charged fermions.

What is lost, though, is the $b-\tau$ unification, i.e. with $\langle(15,2,2)\rangle_{\overline{126}} \neq 0$, $m_b = m_\tau$ at M_{GUT} becomes an accident. However, in the case of type II see-saw, there is a profound connection between $b-\tau$ unification and a large atmospheric mixing angle. The fermionic mass matrices are obtained from (164)

$$\begin{aligned} M_u &= v_{10}^u y_{10} + v_{126}^u y_{126} , \\ M_d &= v_{10}^d y_{10} + v_{126}^d y_{126} , \\ M_e &= v_{10}^d y_{10} - 3v_{126}^d y_{126} , \\ M_{\nu_D} &= v_{10}^u y_{10} - 3v_{126}^u y_{126} , && (166) \\ M_{\nu_R} &= y_{126} \langle (10,1,3) \rangle , && (167) \\ M_{\nu_L} &= y_{126} \langle (\overline{10},3,1) \rangle , && (168) \\ & && (169) \end{aligned}$$

where $\langle(\overline{10},3,1)\rangle \simeq M_W^2/M_{GUT}$ provides a direct (type II) see-saw mass for light neutrinos. The form in (166) is readily understandable, if you notice that $\langle(1,2,2)\rangle$ is a $SU(4)_c$ singlet with $m_q = m_\ell$, and $\langle(15,2,2)\rangle$ is a $SU(4)_c$ adjoint, with $m_\ell = -3m_q$. The vevs of the bidoublets are denoted by v^u and v^d as usual.

Now, suppose that type II dominates, or $M_\nu \propto y_{126} \propto M_e - M_d$, so that

$$M_\nu \propto M_e - M_d \tag{170}$$

Let us now look at the 2nd and 3rd generations first. In the basis of diagonal M_e, and for the small mixing ϵ_{de}

$$M_\nu \propto \begin{pmatrix} m_\mu - m_s & \epsilon_{de} \\ \epsilon_{de} & m_\tau - m_b \end{pmatrix} \tag{171}$$

obviously, large atmospheric mixing can only be obtained for $m_b \simeq m_\tau$ [45].

Exercise 13 *Prove that the above neutrino mass matrix requires $b - \tau$ unification in order to lead to a large mixing angle. Use the fact that the second generation masses are small in comparison with the third generation ones.*

Of course, there was no reason whatsoever to assume type II see-saw. Actually, we should reverse the argument: the experimental fact of $m_b \simeq m_\tau$ at M_{GUT}, and large $\theta_{\rm atm}$ seem to favor the type type II see-saw. It can be shown, in the same approximation of 2-3 generations, that type I cannot dominate: it gives a small $\theta_{\rm atm}$ [50]. This gives hope to disentangle the nature of the see-saw in this theory. As a check, it can be shown that the two types of see-saw are really inequivalent [50].

I wish to stress an important feature of this programme. Since 126 ($\overline{126}$) is invariant under matter parity, R parity remains exact at all energies and thus the lightest supersymmetric particle is stable and a natural candidate for the dark matter.

Mass scales

In SO(10) we have in principle more than one scale above M_W (and Λ_{SUSY}): the GUT scale, the Pati-Salam scale where $SU(4)_c$ is broken, the L-R scale where parity (charge conjugation) is broken, the scales of the breaking of $SU(2)_R$ and $U(1)_{B-L}$. Of course, these may be one and the same scale, as expected with low-energy supersymmetry. This solution is certainly there, since the gauge couplings of the MSSM unify successfully and encourage the single step breaking of SO(10).

Is there any room for intermediate mass scales in SUSY SO(10)? It is certainly appealing to have an intermediate see-saw mass scale M_R, between $10^{12} - 10^{15} GeV$ or so. In the non-renormalizable case, with 16_H and $\overline{16}_H$, this is precisely what happens: $M_R \simeq c M_{GUT}^2 / M_{Pl} \simeq c(10^{13} - 10^{14}) GeV$. In the renormalizable case, with 126_H and $\overline{126}_H$, one needs to perform a renormalization group study using unification constraints. While this is in principle possible, in practice it is hard due to the large number of fields. The stage has recently been set, for all the particle masses were computed [51,52], and the preliminary studies show that the situation may be under control [53]. It is interesting that the existence of intermediate mass scales lowers the GUT scale [51,54], allowing for a possibly observable $d = 6$ proton decay.

Notice that a complete study is basically impossible. In order to perform the running, you need to know particle masses precisely. Now, suppose you stick to the principle of minimal fine-tuning. As an example, you fine-tune the mass of the W and Z in the SM, then you know that the Higgs mass and the fermion masses are at the same scale

$$m_H = \frac{\sqrt{\lambda}}{g} m_W \,, \quad m_f = \frac{y_f}{g} m_W \tag{172}$$

where λ is a ϕ^4 coupling, and y_f an appropriate fermionic Yukawa coupling. Of course, you know the fermion masses in the SM model, and you know $m_H \simeq m_W$.

In an analogous manner, at some large scale m_G a group G is broken and there are usually a number of states that lie at m_G, with masses

$$m_i = \alpha_i m_G \tag{173}$$

where α_i is an approximate dimensionless coupling. Most renormalization group studies typically argue that $\alpha_i \simeq O(1)$ is natural, and rely on that heavily. In the SM, you could then take $m_H \simeq m_W$, $m_f \simeq m_W$; while reasonable for the Higgs, it is nonsense for the fermions (except for the top quark).

In supersymmetry *all* the couplings are of Yukawa type, i.e. self-renormalizable, and thus taking $\alpha_i \simeq O(1)$ may be as wrong as taking all $y_f \simeq O(1)$. While a possibly reasonable approach when trying to get a qualitative idea of a theory, it is clearly unacceptable when a high-precision study of M_{GUT} is called for.

Proton decay

As you know, $d = 6$ proton decay gives $\tau_p(d = 6) \propto M_{GUT}^4$, while $(d = 5)$ gives $\tau_p(d = 5) \propto M_{GUT}^2$. In view of the discussion above, the high-precision determination of τ_p appears almost impossible in SO(10) (and even in SU(5)).

You may wonder if our renormalizable theory makes sense at all. After all, we are ignoring the higher dimensional operators of order $M_{GUT}/M_{Pl} \simeq 10^{-2}-10^{-3}$. If they are present with the coefficients of order one, we can forget almost everything we said about the predictions, especially in the Yukawa sector. However, we actually know that the presence of $1/M_{Pl}$ operators is not automatic (at least not with the coefficients of order 1). Operators of the type (in symbolic notation)

$$O_5^p = \frac{c}{M_{Pl}} 16_F^4 \tag{174}$$

are allowed by SO(10) and they give

$$O_5^p = \frac{c}{M_{Pl}} [(QQQL) + (Q^c Q^c Q^c L^c)] \tag{175}$$

These are the well-known $d = 5$ proton decay operators, and for $c \simeq O(1)$ they give $\tau_p \simeq 10^{23} yr$. Agreement with experiment requires

$$c \leq 10^{-6} \tag{176}$$

Exercise 14 *Hard. Prove the above result. Use the fact that the supersymmetric operator of the type $QQQL$ corresponds to an effective interaction $QL\tilde{Q}\tilde{Q}$ and then use the interactions with gauginos to transform $\tilde{Q}\tilde{Q}$ into QQ in order to create a proton decay operator $QQQL$. It happens at the one loop level.*

Could this be a signal that $1/M_{Pl}$ operators are small in general? Alternatively, you need to understand why just this one is to be so small. It is appealing to assume that this may be generic; if so, neglecting $1/M_{Pl}$ contributions in the study of fermion masses and mixings is fully justified.

6 Summary

These lectures were an attempt to bring the readers up to date in the field of grand unification. After more than thirty years we are still searching for the standard grand unified theory, an analogue of the standard model of electroweak interactions. It does not help that the two main predictions of grand unification, the proton decay and magnetic monopoles were never observed. I wanted to convince you here that the program does have sense and is worth pursuing. There is real hope that the new generation of proton decay experiments may finally lead to its discovery, and the accumulated on neutrino masses and mixings provide a nontrivial check on models.

Acknowledgments

I am very grateful to Meltem Serin-Zeyrek and other organizers of this summer school for giving me the opportunity to present my lectures and for a warm hospitality. The setting, the atmosphere and an excellent school make me wish to be back as soon as possible. I am indebted to my collaborators over the recent years Charan Aulakh, Borut Bajc, Alejandra Melfo and Francesco Vissani for great fun and for teaching me a lot. Last but not least I owe great gratitude to Alejandra Melfo without whom this manuscript would have never been completed.

References

1. A. Strumia and F. Vissani, arXiv:hep-ph/0606054.
2. R. N. Mohapatra and P. B. Pal, World Sci. Lect. Notes Phys. **60** (1998) 1 [World Sci. Lect. Notes Phys. **72** (2004) 1].
3. R. N. Mohapatra, *Berlin, Germany: Springer (1986) 309 P. (Contemporary Physics)*
4. G. G. Ross, *Reading, Usa: Benjamin/cummings (1984) 497 P. (Frontiers In Physics, 60)*
5. J. S. Schwinger, Annals Phys. **2** (1957) 407.
6. H. Georgi and S. L. Glashow, Phys. Rev. Lett. **28**, 1494 (1972).
7. G. 't Hooft, Nucl. Phys. B **79**, 276 (1974).
8. A. M. Polyakov, JETP Lett. **20** (1974) 194 [Pisma Zh. Eksp. Teor. Fiz. **20** (1974) 430].
9. S. L. Glashow, Nucl. Phys. **22**, 579 (1961).
10. S. Weinberg, Phys. Rev. Lett. **19**, 1264 (1967).
11. A. Salam, *Originally printed in *Svartholm: Elementary Particle Theory, Proceedings Of The Nobel Symposium Held 1968 At Lerum, Sweden*, Stockholm 1968, 367-377*
12. H. Georgi and S. L. Glashow, Phys. Rev. Lett. **32**, 438 (1974).
13. P. A. M. Dirac, Phys. Rev. **74**, 817 (1948).
14. S. Dimopoulos, S. Raby and F. Wilczek, Phys. Rev. D **24** (1981) 1681.
15. L. E. Ibáñez and G. G. Ross, Phys. Lett. B **105** (1981) 439.
16. M. B. Einhorn and D. R. Jones, Nucl. Phys. B **196** (1982) 475.
17. W. Marciano and G. Senjanović, Phys. Rev.D **25** (1982) 3092.
18. K. Inoue, A. Kakuto, H. Komatsu and S. Takeshita, Prog. Theor. Phys. **68** (1982) 927 [Erratum-ibid. **70** (1983) 330].
19. L. Alvarez-Gaume, J. Polchinski and M. B. Wise, Nucl. Phys. B **221** (1983) 495.
20. For useful reviews on spinors in SO(2N) see R. N. Mohapatra and B. Sakita, Phys. Rev. D **21** (1980) 1062 and F. Wilczek and A. Zee, Phys. Rev. D **25** (1982) 553. See also P. Nath and R. M. Syed, Nucl. Phys. B **618** (2001) 138 and C. S. Aulakh and A. Girdhar, arXiv:hep-ph/0204097.
21. R. N. Mohapatra, Phys. Rev. Lett. **43**, 893 (1979).
22. P. Nath and P. F. Perez, arXiv:hep-ph/0601023.

23. P. Minkowski, Phys. Lett. B **67** (1977) 421; T. Yanagida, proceedings of the *Workshop on Unified Theories and Baryon Number in the Universe*, Tsukuba, 1979, eds. A. Sawada, A. Sugamoto, S. Glashow, in *Cargese 1979, Proceedings, Quarks and Leptons* (1979); M. Gell-Mann, P. Ramond, R. Slansky, proceedings of the *Supergravity Stony Brook Workshop*, New York, 1979, eds. P. Van Niewenhuizen, D. Freeman, R. Mohapatra, G. Senjanović, Phys. Rev. Lett. **44** (1980) 912.
24. S. Weinberg, Phys. Rev. Lett. **43** (1979) 1566.
25. E. Ma, Phys. Rev. Lett. **81** (1998) 1171 [arXiv:hep-ph/9805219].
26. M. Magg and C. Wetterich, Phys. Lett. B **94** (1980) 61; G. Lazarides, Q. Shafi and C. Wetterich, Nucl. Phys. B **181** (1981) 287; R. N. Mohapatra and G. Senjanović, Phys. Rev. D **23** (1981) 165.
27. R. Foot, H. Lew, X. G. He and G. C. Joshi, Z. Phys. C **44** (1989) 441.
28. J. C. Pati and A. Salam, Phys. Rev. D **10** (1974) 275.
29. R. N. Mohapatra and J. C. Pati, Phys. Rev. D **11** (1975) 2558.
30. G. Senjanović and R. N. Mohapatra, Phys. Rev. D **12** (1975) 1502.
31. G. Senjanović, Nucl. Phys. B **153** (1979) 334.
32. For a review of the seesaw in the context of SO(10), see e.g. G. Senjanović, arXiv:hep-ph/0501244.
33. I. Dorsner and P. F. Perez, Nucl. Phys. B **723** (2005) 53 [arXiv:hep-ph/0504276].
34. B. Bajc and G. Senjanović, arXiv:hep-ph/0612029.
35. J. R. Ellis and M. K. Gaillard, Phys. Lett. B **88** (1979) 315.
36. B. Bajc, M. Nemevsek and G. Senjanovic, arXiv:hep-ph/0703080.
37. S. Dimopoulos, S. Raby and F. Wilczek, Phys. Rev. D **24** (1981) 1681.
38. L. E. Ibanez and G. G. Ross, Phys. Lett. B **105** (1981) 439.
39. M. B. Einhorn and D. R. T. Jones, Nucl. Phys. B **196** (1982) 475.
40. W. J. Marciano and G. Senjanović, Phys. Rev. D **25** (1982) 3092.
41. B. Bajc, M. Nemevšek and G. Senjanović, work in progress.
42. R. N. Mohapatra, *Phys. Rev.* **D 34**, 3457 (1986). A. Font, L. E. Ibáñez and F. Quevedo, *Phys. Lett.* **B 228**, 79 (1989). S. P. Martin, *Phys. Rev.* **D 46**, 2769 (1992). C. S. Aulakh, K. Benakli and G. Senjanović, Phys. Rev. Lett. **79** (1997) 2188. C. S. Aulakh, A. Melfo and G. Senjanović, Phys. Rev. D **57**, 4174 (1998); [arXiv:hep-ph/9707256]. C. S. Aulakh, A. Melfo, A. Rašin and G. Senjanović, Phys. Rev. D **58**, 115007 (1998) [arXiv:hep-ph/9712551]. C. S. Aulakh, A. Melfo, A. Rašin and G. Senjanović, Phys. Lett. B **459** (1999) 557. [arXiv:hep-ph/9902409].
43. C. S. Aulakh, K. Benakli and G. Senjanović, Phys. Rev. Lett. **79** (1997) 2188. C. S. Aulakh, A. Melfo and G. Senjanović, Phys. Rev. D **57**, 4174 (1998). C. S. Aulakh, A. Melfo, A. Rašin and G. Senjanović, Phys. Lett. B **459** (1999) 557.
44. C. S. Aulakh, B. Bajc, A. Melfo, A. Rašin and G. Senjanović, Nucl. Phys. B **597** (2001) 89.
45. B. Bajc, G. Senjanović and F. Vissani, Phys. Rev. Lett. **90** (2003) 051802.
46. K. S. Babu and R. N. Mohapatra, Phys. Rev. Lett. **70**, 2845 (1993).
47. E. Witten, Phys. Lett. B **91** (1980) 81.
48. See e.g. D. Chang, R. N. Mohapatra, J. Gipson, R. E. Marshak and M. K. Parida, Phys. Rev. D **31** (1985) 1718.
49. G. Dvali and A. Vilenkin, arXiv:hep-th/0304043. G. Dvali, arXiv:hep-th/0410286.

50. B. Bajc, G. Senjanović and F. Vissani, Phys. Rev. D **70** (2004) 093002 [arXiv:hep-ph/0402140].
51. B. Bajc, A. Melfo, G. Senjanović and F. Vissani, Phys. Rev. D **70** (2004) 035007.
52. T. Fukuyama, A. Ilakovac, T. Kikuchi, S. Meljanac and N. Okada, arXiv:hep-ph/0401213. C. S. Aulakh and A. Girdhar, arXiv:hep-ph/0204097.
53. C. S. Aulakh and A. Girdhar, arXiv:hep-ph/0405074.
54. H. S. Goh, R. N. Mohapatra and S. Nasri, arXiv:hep-ph/0408139.
55. K. S. Babu and C. Macesanu, Phys. Rev. D **72** (2005) 115003 [arXiv:hep-ph/0505200]. S. Bertolini, M. Frigerio and M. Malinsky, Phys. Rev. D **70**, 095002 (2004) [arXiv:hep-ph/0406117].
56. H. Arason, D. J. Castano, E. J. Piard and P. Ramond, Phys. Rev. D **47** (1993) 232 [arXiv:hep-ph/9204225].
57. C. S. Aulakh and R. N. Mohapatra, Phys. Rev. D **28** (1983) 217.
58. T. E. Clark, T. K. Kuo and N. Nakagawa, Phys. Lett. B **115** (1982) 26.
59. D. Chang, R. N. Mohapatra and M. K. Parida, Phys. Rev. D **30** (1984) 1052. X. G. He and S. Meljanac, Phys. Rev. D **41** (1990) 1620. D. G. Lee, Phys. Rev. D **49** (1994) 1417. D. G. Lee and R. N. Mohapatra, Phys. Rev. D **51** (1995) 1353.
60. C. S. Aulakh, B. Bajc, A. Melfo, G. Senjanović and F. Vissani, Phys. Lett. B **588**, 196 (2004).
61. B. Bajc, A. Melfo, G. Senjanovic and F. Vissani, Phys. Rev. D **73** (2006) 055001 [arXiv:hep-ph/0510139].

Neutrino Physics and Astrophysics: Highlights

Henry T. Wong

Institute of Physics, Academia Sinica, Taipei 11529, Taiwan
htwong@phys.sinica.edu.tw

Abstracts
This article presents an overview of neutrino physics research, with highlights on the physics goals, results and interpretations of the current neutrino experiments and future directions and program.

1 Introduction

One of the major breakthrough in elementary particle physics in the past decade is the evidence of neutrino masses and mixings through the studies of neutrino oscillations. It offers the first case for the need to extend the much-cherished Standard Model. Several important questions are raised, and there are intense world-wide efforts to pursue the next-generation of neutrino experiments to address them.

The objective of this article is to "set the stage" for students and researchers not in the field by summarizing the key ingredients and highlights of the goals, status and future directions in neutrino physics. It is not meant to be a comprehensive lecture or detailed review article. Interested readers can pursue the details via the listed references in the Review of Particle Physics [1] textbooks [2], conference proceedings [3] and Web-links [4].

2 Neutrino Physics: WHY

Neutrino exists – and exists in large quantities in the Universe, comparable in number density to the photon. It is known that there are three flavor of light neutrino coupled via weak interaction to the Z gauge boson. Similar to the quark sector, the flavor (or interaction) eigenstates ν_l is a linear combination of the mass eigenstates ν_i via a mixing matrix U, usually called the PMNS matrix (after Pontecorvo-Maki-Nakagawa-Sakata). Denoting the flavor and mass eigenstates respectively by

$$\Psi \equiv \begin{pmatrix} \nu_e \\ \nu_\mu \\ \nu_\tau \end{pmatrix} \quad \& \quad \phi \equiv \begin{pmatrix} \nu_1 \\ \nu_2 \\ \nu_3 \end{pmatrix} , \tag{1}$$

one have

$$\Psi = U \phi . \tag{2}$$

Unitarity constraints imply that U can be described four independent variables which are usually parameterized as:

$$U \equiv U (\theta_{12} , \theta_{23} , \theta_{13} , \delta) \tag{3}$$

where θ_{ij} are the mixing angles between ν_i and ν_j, and δ is a phase that characterizes possible CP-violation. Being electrically-neutral, the neutrinos can be "Dirac" or "Majorana" particles, identified by whether the neutrinos and anti-neutrinos are different or the same, respectively. For Majorana neutrinos, a diagonal matrix

$$U_{Maj} = diag (1 , e^{i\alpha} , e^{i\beta}) \tag{4}$$

should be added such that the mixing matrix becomes $U \rightarrow UU_{Maj}$.

The "road map" in neutrino physics research can be summarized in Fig. 1. In field theory language, the experimentally measurable neutrino properties can be encoded in a "neutrino mass term" $L(m_\nu)$ in the Lagrangian. Standard Model sets this to be identically zero, but without any compelling reasons – in contrast to the massless-ness of the photons being dictated by gauge invariance. The detailed structures and values of this term can reveal much about the Grand Unified Theories.

The mass term can be expressed either as:

$$L(m_\nu) = \bar{\Psi}_R \cdot M_{3X3} \cdot \Psi_L \tag{5}$$

in the flavor basis or as

$$L(m_\nu) = \bar{\phi}_R \cdot M_D \cdot \phi_L \tag{6}$$

in the mass basis, where the 3×3 matrix M_{3X3} is related to the diagonalized mass matrix M_D by

$$M_{3X3} = U^T \cdot M_D \cdot U. \tag{7}$$

The fundamental intrinsic neutrino properties of great interest are the "neutrino masses" m_i of ν_i (for $i = 1, 2, 3$) given in

$$U_D = diag (m_1 , m_2 , m_3) \tag{8}$$

and the "neutrino mixings" due to individual elements U_{li} of U. The m_i and U_{li} are experimentally measurable, from which M_{3X3} can be constructed. The high energy new physics scale as well as parameters for symmetry breaking and

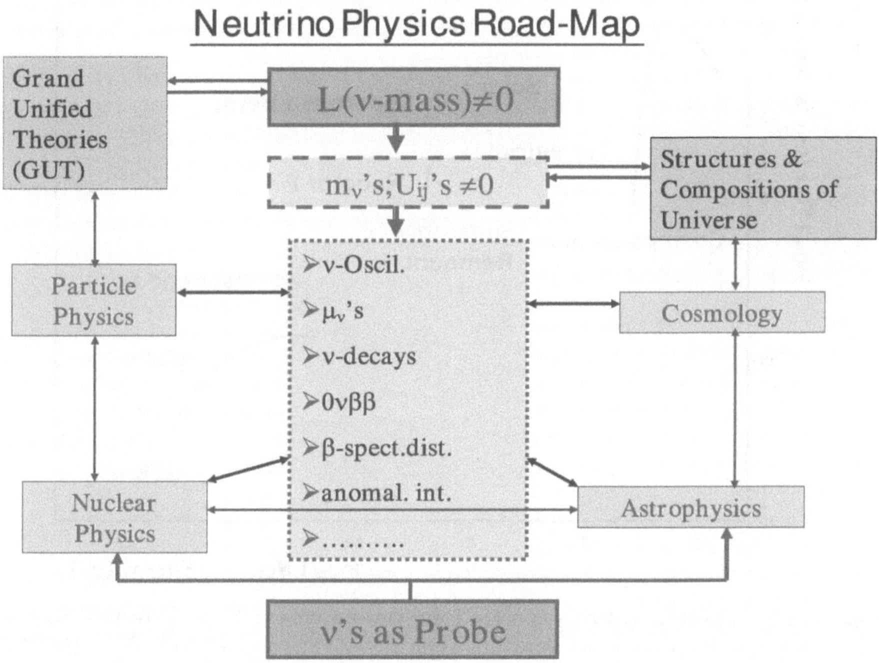

Fig. 1. A schematic chart showing the relationships between different domains in neutrino physics research.

mass generation are expected to be leave their imprints in M_{3X3}. Accordingly, the pursuit of neutrino masses and mixings will lead to a knowledge of M_{3X3} which will shed light to the deep-hidden secrets of nature.

At the large length-scale frontier, neutrino mass is related to the composition and structural evolution of the Universe. The combined neutrino and cosmology data [5] indicate that our Universe is at critical density and that neutrinos constitute to at least 0.1% of this density, comparable to the fraction shared by visible matter.

The study of neutrino physics and the implications of the results connect many disciplines together, from particle physics to nuclear physics to astrophysics to cosmology. Experimentally, the probing of $L(m_\nu)$ is carried out by studying various processes related to neutrino masses and mixings, such as direct mass measurement through the distortion of β-spectra, neutrinoless double beta decays, neutrino oscillations, neutrino magnetic moments, neutrino decays ..... and so on. These investigations are realized by a wide spectrum of experimental techniques spanned over many decades of energy scale with different neutrino sources. The expected neutrino spectrum due to terrestrial and astrophysical sources are shown in Fig. 2.

In addition, neutrino has been used as probe (as "beam" from accelerators and reactors and even astrophysical sources) to study electroweak physics,

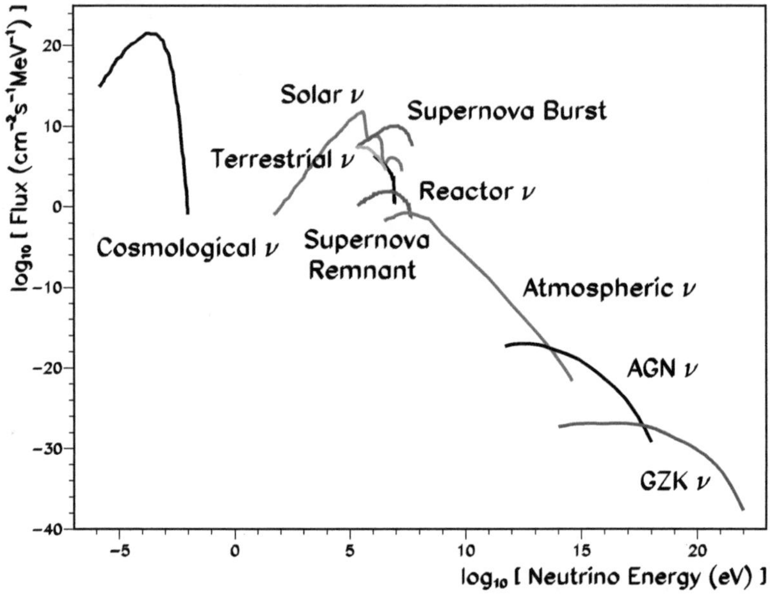

Fig. 2. The expected neutrino spectra from various celestial and terrestrial sources. Neutrinos from man-made accelerators, typically at the range of 10 MeV to 100 GeV, are not shown here since accelerator parameters differ. The detectable window is also shown.

QCD, structure function physics, nuclear physics, and to provide otherwise inaccessible information on the interior of stars and supernovae.

Neutrino interactions are characterized by cross-sections at the weak scale (100 fb at 100 GeV to $<10^{-4}$ fb at 1 MeV). As an illustration, the mean free path in water for $\bar{\nu_e}$ from power reactors at the typical energy of 2 MeV is 250 light years! The central challenge to neutrino experiments is therefore on how to beat this small cross section. The typical solution is to deploy massive detectors. Consequently the experimental hurdles become how to keep the cost and background manageable. These pose great challenges and demand to the creativity – and often courage – of experimentalists.

Neutrino Physics: NOW

After half a century of ingenious experiments since the experimental discovery of the neutrinos by Cowan and Reines [6], there are several results which indicate compellingly the existence of neutrino masses [7, 8], and hence the necessity for the extension of the Standard Model. All these results are based on experimental studies of neutrino oscillation, a quantum-mechanical interference effect which allows neutrino to transform from one flavor eigenstate to another as it traverses in space.

Using the two-flavor formulation of the two lighter families as illustration, Eq. 2 can be written as

$$\begin{pmatrix} \nu_e \\ \nu_\mu \end{pmatrix} = \begin{pmatrix} cos\ \theta_{12} & sin\ \theta_{12} \\ -sin\ \theta_{12} & cos\ \theta_{12} \end{pmatrix} \begin{pmatrix} \nu_1 \\ \nu_2 \end{pmatrix} . \tag{9}$$

The probability of ν_e at energy E_ν transforming to ν_μ after traversing a distance L is given by

$$P(\nu_e \rightarrow \nu_\mu) = sin^2 2\theta\ sin^2(\frac{\Delta m^2_{12}\ L}{E_\nu}) \tag{10}$$

where $\Delta m^2_{12} = m^2_2 - m^2_1$ is the mass-squared difference between the two mass eigenstates. The (E_ν, L) parameters are select-able in individual experiments such that different regions of $(\Delta m^2, \theta)$ are probed. In the presence of matter, a resonance "MSW" effect [9] can take place when the condition

$$\sqrt{2} \cdot G_F \cdot \rho_e = \frac{\Delta m^2}{2E_\nu} \cdot \cos 2\theta \tag{11}$$

are satisfied, where G_F is the Fermi constant and ρ_e is the electron density in matter. The effective mixings at resonance are maximal even though the vacuum mixing angle θ can be small. The origins are due to the physics that (ν_μ, ν_τ) only interact via neutral currents with the electrons while ν_e can have both neutral and charged-current interactions. A summary of the results of neutrino oscillation experiments is shown in Fig. 3.

Our current knowledge of the mixing matrix U and of Δm^2 is derived from positive evidence of neutrino masses and mixings from several experiments using different neutrino sources:

2.1 Atmospheric and Accelerator Neutrinos

Data from the Super-Kamiokande experiments [10], supported by other experiments (MACRO, SOUDAN), indicate a smaller $(\nu_\mu + \bar{\nu_\mu})/(\nu_e + \bar{\nu_e})$ ratio than would be expected from propagation models of cosmic-ray showering. There is a statistically strong dependence of the deficit with the zenith angle of the neutrinos, which can be translated to a dependence with L/E_ν, a "smoking gun" evidence for neutrino oscillations. The data supports a scenario of $\nu_\mu \rightarrow \nu_\tau$ oscillation, rather than ν_ν disappearing into sterile neutrinos.

The allowed region is indicated by "SuperK" in the $(\Delta m^2, \theta)$ plot of Fig. 3, with

$$\begin{aligned} \Delta m^2_{atm} &= (1.9 - 3.0) \times 10^{-3}\ \mathrm{eV}^2 \\ sin^2 2\theta_{atm} &> 0.9 \end{aligned} \tag{12}$$

Long baseline neutrino oscillation experiment K2K [11] compared the flux of an accelerator ν_μ beam close to its production point at the KEK Laboratory to that measured at the Super-Kamiokande detector after the beam

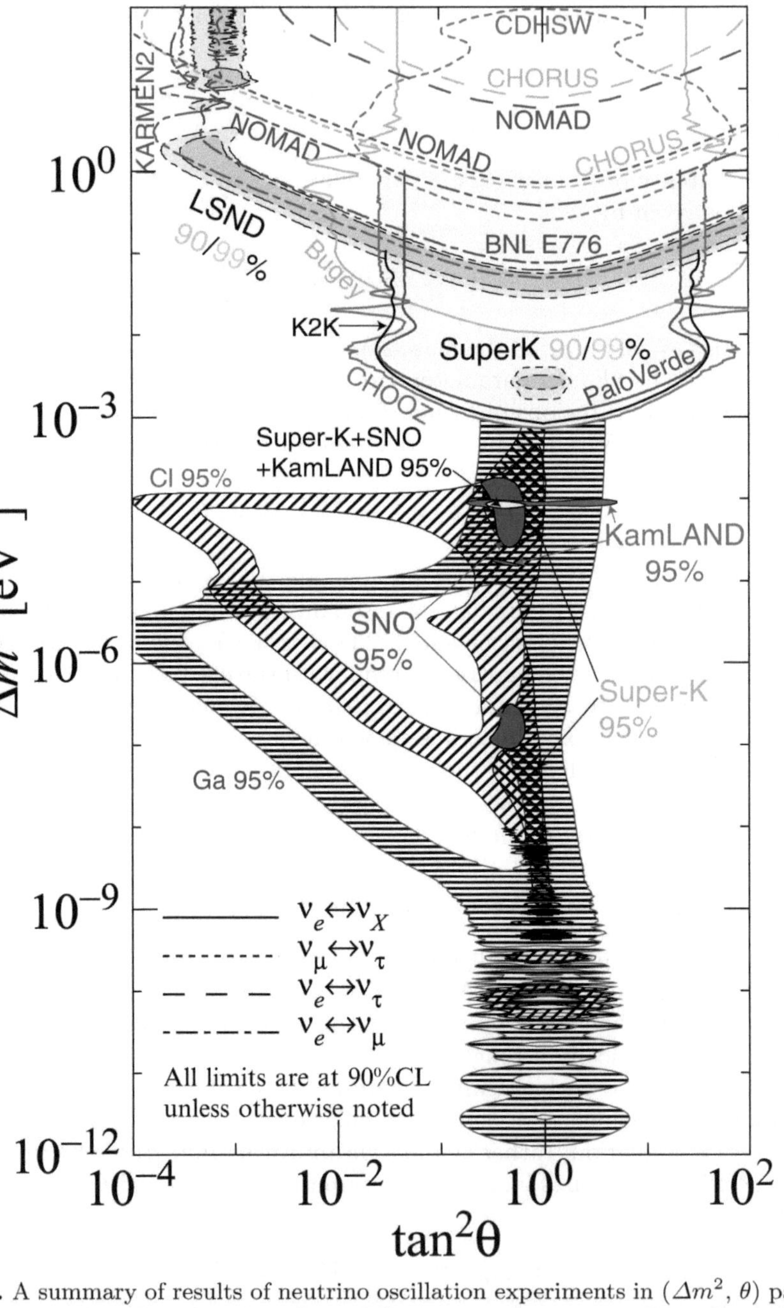

Fig. 3. A summary of results of neutrino oscillation experiments in $(\Delta m^2, \theta)$ parameter space. The allowed regions from different experiments are shown by various shading. Adapted from Ref. [7].

traversed a distance of 250 km. Disappearance of the ν_μ was observed at a level compatible with the $(\Delta m^2_{23}, \theta_{23})$ measurements of the atmospheric neutrino experiments.

2.2 Solar and Reactor Neutrinos

All previous solar neutrino experiments (Homestake, Kamiokande, GALLEX, SAGE, Super-Kamiokande) observed solar neutrino flux less than the predictions of Standard Solar Model [12]. The deficit is different among the experiments, suggesting an energy dependence of the effect.

The SNO experiment [13] in Canada, with its deuteron target in the form heavy water, measured the solar neutrino flux in three different channels. As depicted in Fig. 4, the flavor-dependent elastic scattering (ES) and charged-current (CC) channels shows deficit while the flavor-independent neutral-current channel agrees perfectly with the predictions from the standard Solar Model [12, 14]. These represent compelling evidence that neutrinos are produced in the correct amount by nuclear fusion in the interior of the Sun, and that some ν_e's produced are converted to another active neutrino flavors on its way to Earth. The "Solar Neutrino Problem", which has been a major puzzle in basic science for thirty years, is thus solved.

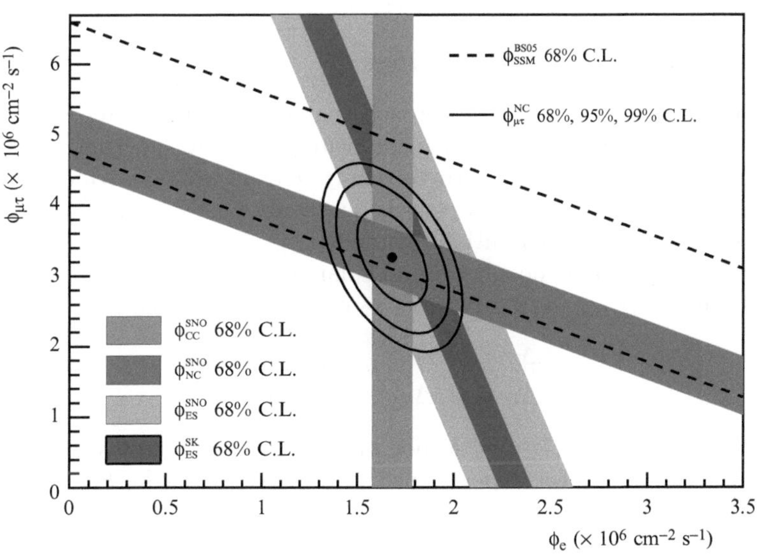

Fig. 4. Results from the SNO experiment showing the complementary regions of sensitivities from the charged-currents (CC), neutral currents (NC) and electron scattering (ES) measurements. Predictions by the Standard Solar Model are the region between the dotted lines. Adapted from Ref. [13].

The combined solar neutrino data can be explained by neutrino oscillation in matter and select the "Large Mixing Angle" (LMA) region as labeled by "SNO" in Fig. 3 as the preferred solution.

The KamLAND [15] experiment in Kamioka, Japan is sensitive to the $\bar{\nu_e}$ emitted from the many nuclear power reactors in the vicinity, with an average distance of 180 km. An energy-dependent deficit of the $\bar{\nu_e}$ flux was observed. When interpreted as neutrino oscillation in vacuum, the allowed parameter space as indicated by "KamLAND" in Fig. 3 has excellent overlap with the solar neutrino LMA region and significantly enhance the sensitivity to Δm^2. The combined fit of the solar and reactor neutrino data gives:

$$\begin{aligned} \Delta m^2_\odot &= 8.0 \pm^{0.6}_{0.4} \times 10^{-5}\ \mathrm{eV}^2 \\ tan^2\theta_\odot &= 0.45 \pm^{0.09}_{0.07} \end{aligned} \tag{13}$$

2.3 LSND Anomaly:

The LSND experiment [16] with accelerator neutrinos reported unexpected excess of ν_e in a $\nu_\mu + \bar{\nu_\mu}$ beam, which can be explained by $\nu_\mu \rightarrow \nu_e$ oscillation at the relative large Δm^2 and small θ region as denoted by "LSND" in Fig. 3. The results are yet to be reproduced (or totally excluded) by other experiments. The MiniBoone experiment, currently under data taking and analysis, will confirm or refute these anomalous results.

2.4 Neutrino Masses and Mixings

If one takes the conservative approach that the LSND results must be confirmed by an independent experiment before they are incorporated into the theoretical framework, then a three-family scheme with neutrino masses and mixings is adequate. Otherwise, sterile neutrinos must be added and the Standard Model must be substantially extended.

Accordingly, the atmospheric and solar neutrino results given in Eqs. 12 and 13 are due to $\nu_2 \leftrightarrow \nu_3$ and $\nu_1 \leftrightarrow \nu_2$ oscillations, respectively:

$$\Delta m^2_{23} = \Delta m^2_{atm}\ ;\ \theta_{23} = \theta_{atm} \tag{14}$$

$$\Delta m^2_{12} = \Delta m^2_\odot\ ;\ \theta_{12} = \theta_\odot \tag{15}$$

The oscillation and $\nu_1 \leftrightarrow \nu_3$ is not observed experimentally yet, and only upper limit from reactor experiment CHOOZ [17] exists for θ_{13}:

$$\begin{aligned} \Delta m^2_{13} \ \sim\ \Delta m^2_{23} &= (1.9 - 3.0) \times 10^{-3}\ \mathrm{eV}^2 \\ sin^2\theta_{13} &< 0.05 \end{aligned} \tag{16}$$

Taking $\theta_{12} = 45^o$, the mixing matrix U in Eq. 3 has the structure:

$$U \sim \begin{pmatrix} cos\theta_{12} & sin\theta_{12} & e^{i\delta} sin\theta_{13} \\ -sin\theta_{12}/\sqrt{2} & cos\theta_{12}/\sqrt{2} & 1/\sqrt{2} \\ sin\theta_{12}/\sqrt{2} & -cos\theta_{12}/\sqrt{2} & 1/\sqrt{2} \end{pmatrix}, \tag{17}$$

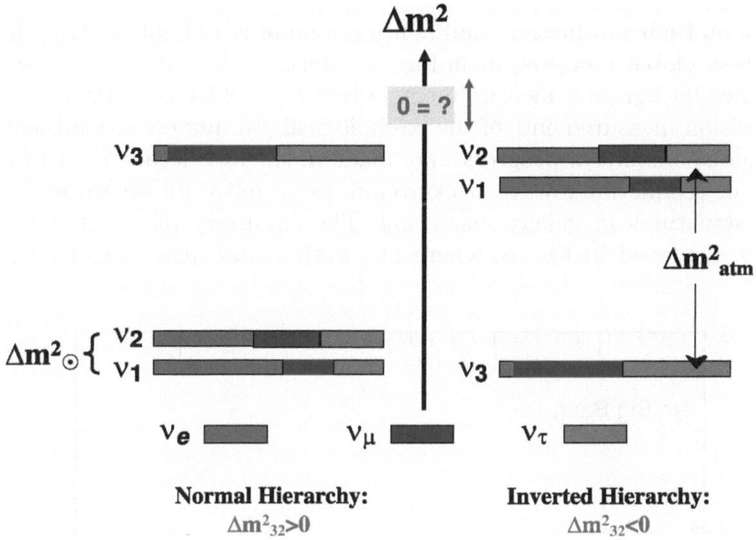

Fig. 5. The favored neutrino mass spectra to explain the various positive results from neutrino oscillation experiments.

where $\theta_{12} \sim 34^o$. The mixing angle θ_{13} and the CP-violating phase δ remain unknown. The matrix U_{Maj} is diagonal and hence is not sensitive to oscillation experiments. An important feature is that while the formalism are almost identical, the structure of the mixing matrix in the neutrino sector is very different from that in the quark sector. In particular, cases where $\theta_{23} = 45^o$ or $\theta_{13} = 0$ exactly may imply symmetry principles in nature yet to be discovered.

Our present knowledge of the neutrino mass spectrum is depicted schematically in Fig. 5. The color-codes denote the compositions of the various mass eigenstates. Depending on which is the eigenstate with the minimum mass, there are two possibilities: the normal and the invert hierarchies. Since the oscillation results only provides information on Δm^2's, the absolute scale of m_i remains unknown. It can be "hierarchical", in which $m_i(max) \sim \sqrt{\Delta m^2_{23}} \sim$ 45 meV, or "quasi-degenerate", where $m_1 \sim m_2 \sim m_3$. Existing β-spectrum measurements give upper bound of $m_i < 2$ eV for the individual neutrino masses.

3 Neutrinos in Cosmology

There is a long tradition of interplay between neutrino physics and cosmology – that is, physics at the smallest and largest scale. The standard model of primordial nucleosynthesis [18] is a pillar of the Big-Bang Cosmology. Comparisons of the observed baryon-to-photon ratio (η) in the Universe with the abundance of ^{4}He, D, ^{3}He and ^{7}Li can place constraints on the degrees of freedom on their production, and hence the number of light neutrino families. The latest global analysis, including a determination of η from the cosmic microwave background measurements, gives $N_\nu = 3.24 \pm 1.2$ [19].

Precision measurements of the cosmological parameters is a subject where tremendous amount of progress were made in the last decade [5], through the studies of cosmic microwave background, large red-shift supernovae, as well as the structures in galaxy clustering. The summary plot in the $(\Omega_m, \Omega_\Lambda)$ plane is displayed in Fig. 6, where Ω_m is the total mass density, while Ω_Λ

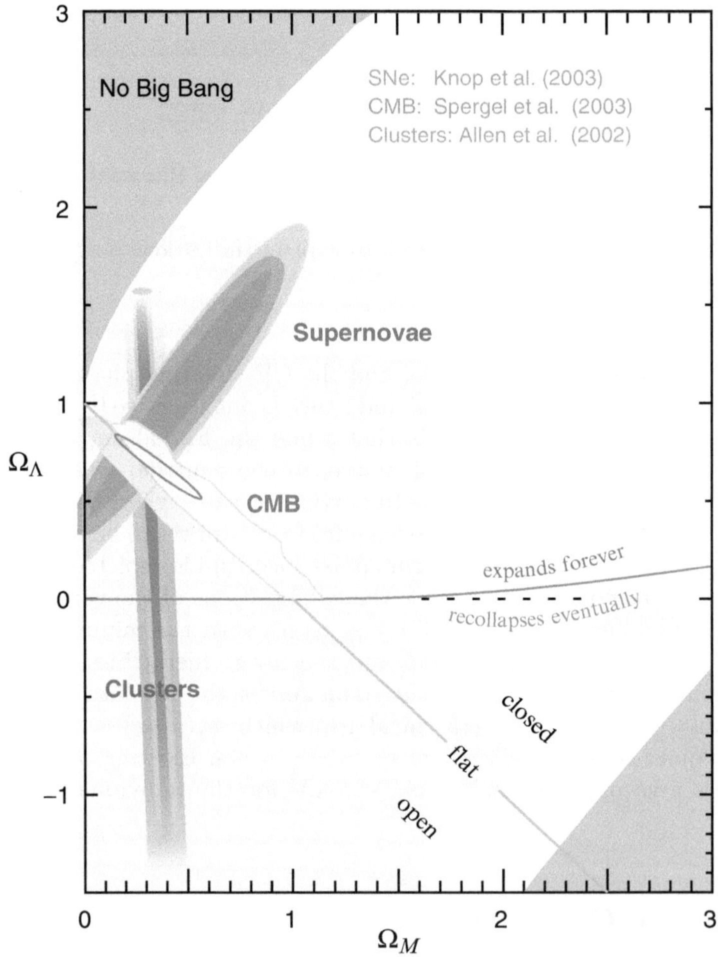

Fig. 6. Summary plot showing the region of sensitivities of the various cosmology measurements in the $(\Omega_m, \Omega_\Lambda)$ plane. Adapted from Ref. [5].

is the cosmological constant. The best-fit allowed region is consistent with a flat Universe ($\Omega_m + \Omega_\Lambda \approx 1$) with about $\Omega \sim 25\%$ in the form of known or yet-unobserved matter, and the remaining $\Omega \sim 75\%$ as "Dark Energy".

The total neutrino mass density affects the scale of structure formation on the galaxy clusters. Combined global analyses of the different probes give rise to upper bound on total neutrino mass of $\sum m_i < 0.17$ eV at 95% confidence level [20]. Coupling with results from neutrino oscillation experiments, lower bound of neutrino mass density of $\Omega_\nu > 0.001$ can be placed.

4 Neutrino Physics: FUTURE

The main thrust of neutrino physics will be to perform precision studies on the already measured parameters, to uncover the remaining parameters (θ_{13}, δ) in U, to establish an mass scale for m_i, to differentiate the two possible mass hierarchies, and to identify neutrinos as being Dirac or Majorana particles. In addition, among the various neutrino sources depicted in Fig. 2, only a relatively small window from ~1 MeV to ~100 GeV is detectable by present techniques. Research efforts towards discovering and studying and devising new neutrino sources, interaction channels as well as detection mechanisms remain crucial and complementary. The future of neutrino experiments will therefore evolve along the various directions:

4.1 Oscillation Parameters:

There is a running experiment with accelerator neutrinos from the Fermilab NuMI beam to the MINOS detector 730 km away. First results [21] are consistent with interpretations of neutrino oscillation at the atmospheric parameter space of Eq. 12. The CERN neutrino beam line is under construction, and will be received by the ICARUS and OPERA experiments at Gran Sasso, also 730 km away. Another approved project is the T2K experiment, where an intense neutrino beam in Tokai, Japan is being built to be sent to the Super-Kamiokande detector 295 km away in an "off-axis" mode. An alternative project NOνA is being pursued using the Fermilab NuMI beam. The goals of these experiments will be to improve on the $(\Delta m^2_{23}, \theta_{23})$ measurements, to detect ν_τ appearance explicitly, to confirm $\nu_\mu \to \nu_\tau$ oscillations, and to measure θ_{13} by searching for $\nu_\mu \to \nu_e$ oscillations, and to possibly probe the neutrino mass hierarchies.

If θ_{13} is finite and the mixing is at least at the 1% level, the CP violating effects in the neutrino sector can be studied, for instance, by comparing the difference between oscillation probabilities P($\nu_e \to \nu_\mu$) and P($\bar{\nu_e} \to \bar{\nu}_\mu$). To pursue this program, one needs intense neutrino beams and Mega-ton size detector. There are intense efforts in devising neutrino "factory", either with improved conventional proton-on-target techniques, or as decay products from muons or radioactive β-decay isotopes produced in accelerators.

Complementary to accelerator-based efforts, there are several reactor-based projects with goals to search for θ_{13} in reactor with about 1 km baseline, by looking for $\bar{\nu_e}$ disappearing through the comparison of the neutrino spectra at this distance to those at production point.

4.2 Intrinsic Neutrino Properties:

Distortions of the end-point of the β-decay spectra provides direct measurements of the neutrino masses m_i's and mixings $U_{ei}U_{ei}^*$. The KATRIN experiment under construction will push the sensitivity of to 0.2 eV, and therefore can probe the scenario with quasi-degenerate neutrino mass eigenstates, as illustrated in Fig. 7. This sensitivity cannot resolve individual m_i and only the effective quantity "electron neutrino mass"

$$m_{\bar{\nu_e}}^2 = \sum_i |U_{ei}|^2 m_i^2 \tag{18}$$

can be determined or constrained.

Neutrinoless Double Beta Decay is the most promising avenue to identify the Dirac or Majorana nature of neutrinos, since this process is possible only for Majorana neutrinos. The decay rates depend on an "effective" mass parameter:

$$\langle m_{\beta\beta} \rangle^2 = \left| \; |U_{e1}|^2 m_1 + e^{i\alpha} |U_{e2}|^2 m_2 + e^{i\beta} |U_{e3}|^2 m_3 \; \right|^2 \tag{19}$$

and hence are sensitive to the absolute scale of $\mathrm{m_i^2}$ rather then $\Delta \mathrm{m_{ij}^2}$. As depicted in Fig. 7, when coupled with the input from oscillation results, this processes can help to distinguish the two possible mass hierarchies. Many R&D projects on various candidate isotopes are being pursued with the eventual goals of achieving the interesting range of $\langle m_{\beta\beta} \rangle \sim 0.1 - 0.01$ eV, which covers the case of neutrino mass spectrum having the inverted hierarchy, as shown in Fig. 7.

Finite neutrino masses typically give rise to anomalous neutrino interactions not accounted for by the Standard Model. The primary example is the possible couplings between neutrinos and photons via their spin components, manifested as neutrino magnetic moments and neutrino radiative decays [22]. Besides the model-dependent astrophysical bounds, the most stringent laboratory direct limits come from reactor neutrino experiments [23], at $\mu_\nu(\bar{\nu}_e) < 7.4 \times 10^{-11}$ μ_B. Bounds are also placed for the ν_e-properties [24] as well as on axion emissions [25] from the reactor experiment.

4.3 Detection of Weak/Rare Signals:

The KamLAND experiment, with its excellent sensitivity to $\bar{\nu_e}$, has recently made the first observation of the terrestrial "geoneutrino" [26] which are by-products of the β-decays due to radioactivity in the Earth's crust, predominantly from the ^{238}U and ^{232}Th series. The various big underground

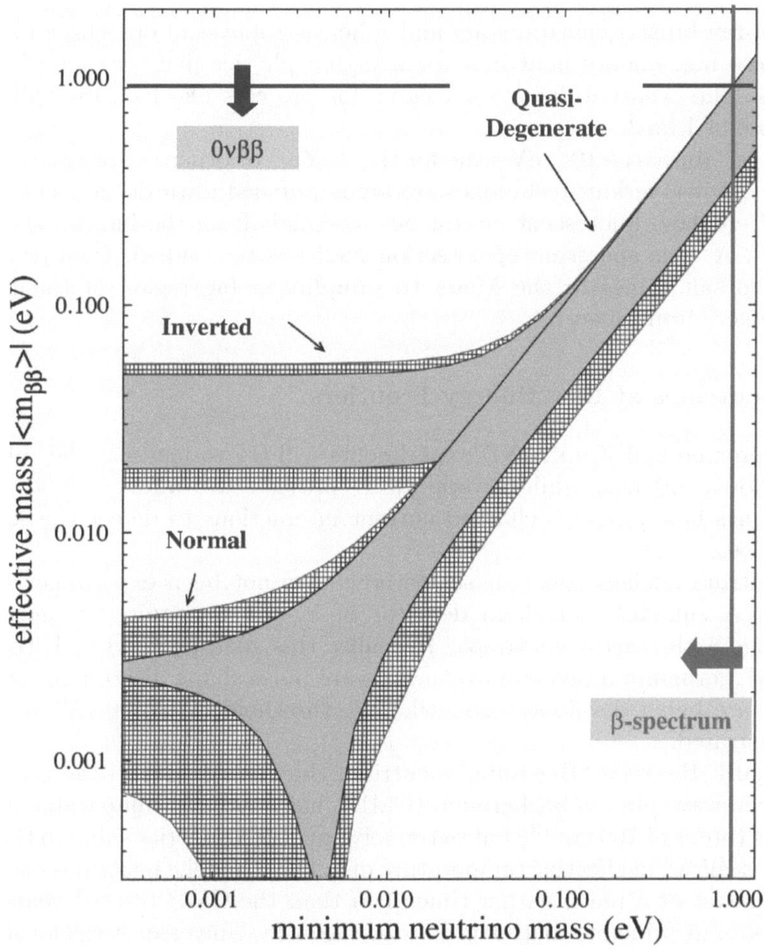

Fig. 7. Sensitivity of the double beta decay $\langle m_{\beta\beta} \rangle$ parameter versus the minimum neutrino mass. The allowed region from oscillation experiments are shown by the bands. The inner and outer bands are due to the best-fit oscillation parameters and with 1σ errors included, respectively. Adapted from Ref. [8].

experiments are sensitive to neutrinos from supernovae, with the expectation of detecting thousands of events from the next supernova, as compared to the 20 events from SN1987a.

4.4 High Energy Astrophysical and Cosmological Neutrinos:

There are several "neutrino telescope" experiments [3] (Lake Baikal, AMANDA, NESTOR, ANTARES, IceCube) based on the water or ice Cerenkov detection techniques. Their scientific goals are (a) to identify and understand the high-energy (10^{12} to 10^{15} eV) neutrino sources from active galactic nuclei, gamma-ray bursts, neutron stars and other astrophysical objects, and (b) to use these high-energy neutrinos for neutrino physics like very long baseline studies. The eventual detection volume for projects like IceCube will be on the scale of 1 km^3.

To get above the 10^{18} eV scale for the "GZK" or other neutrinos of cosmological origins, various techniques are being pursued towards the detection of radio Cerenkov, fluorescent, and acoustic signals from the interaction showers [27]. A wide spectrum of detection media is be studied, from ice to sea water to salt mines to the Moon to sampling a big region of the Earth's atmosphere from Space.

4.5 Neutrinos at Low Energy Frontiers:

The Borexino and KamLAND experiments will try to measure the sub-MeV solar ^{7}Be neutrinos, while several R&D projects are under way to devise techniques based on ν_eN-charged current interactions to detect the solar pp neutrinos.

Neutrino nucleus coherent scatterings have not been experimentally observed. A sub-KeV threshold detector of kg-size is required to meet this challenge with reactor neutrinos. To realize this goal, prototype "Ultra-Low-Energy" germanium detectors with software pulse shape discrimination techniques are being developed [28], where a threshold of O(100 eV) has been demonstrated.

Finally, the relic "Big Bang" neutrino, the counterpart to the 2.7 K cosmic microwave photon background (CMB), has large and comparable number density (order of 100 cm^{-3}) but extremely small cross sections due to the MeV energy scale at an effective temperature of 1.9 K. The relic neutrinos decouples from matter at a much earlier time (1 s) than the CMB (3×10^5 years), and hence are, in principle, better probes to the early Universe. A demonstration of its existence and a measurement of its density is a subject of extraordinary importance. Though there is no realistic proposals on how to detect them, it follows the traditions of offering a highly rewarding challenge to and pushing the ingenuity of neutrino experimentalists.

5 Outlook

Neutrino physics and astrophysics will remain a central subject in experimental particle physics in the coming decades and beyond. The structures of the neutrino mass spectrum and the mixing matrix will be studied more thoroughly, as will be the intrinsic neutrino properties as well as the high and low energy frontiers. There are much room for ground-breaking technical innovations - as well as potentials for *further* surprises.

Acknowledgments

The author is grateful to the hospitality of the ISSSMB06 organizers for the thoughtful arrangement and the superb setting of the summer school, and to the participants whose enthusiasm makes it so memorable. They are to be congratulated for this success, and are encouraged to turn this into a regular national, regional, or even international event. The work is supported by contract 94-2112-M-001-028 from the National Science Council, Taiwan.

References

1. "Review of Particle Physics", Particle Data Group, http://pdg.lbl.gov/, J. Phys. **G 33** (2006).
2. For textbooks, see, for example:
   "Neutrino Physics", ed. K. Winter, Cambridge University Press (1991);
   "Physics of Massive Neutrinos", 2nd Edition, F. Boehm and P. Vogel, Cambridge University Press (1992);
   "Neutrino Physics", K. Zuber, Inst. of Phys. Press (2004).
3. For latest status, see, for example:
   "Proc. of the XXIst Int. Conf. on Neutrino Phys. & Astrophysics", eds. J. Dumarchez, Th. Patzak and F. Vannucci, Nucl. Phys. **B** (Procs. Suppl.) **143** (2005).
4. For Web-portals, click, for example:
   "Neutrino Oscillation Industry" – http://neutrinooscillation.org/;
   "Neutrino Unbound" – http://www.nu.to.infn.it/.
5. O. Lahav and A.R. Liddle, in Ref. [1], J. Phys. **G 33**, 224 (2006), and references therein.
6. F. Reines and C.L. Cowan, Phys. Rev. **90**, 492 (1953).
7. B. Kayser, in Ref. [1], J. Phys. **G 33**, 156 (2006), and references therein.
8. P. Vogel and A. Piepke, in Ref. [1], J. Phys. **G 33**, 479 (2006), and references therein.
9. L. Wolfenstein, Phys. Rev. **D 17**, 2369 (1978); S.P. Mikheyev and A. Yu. Smirnov, Sov. J. Nucl. Phys. **42**, 1441 (1985).
10. Y. Ashie et al., Super-Kamiokande Coll., Phys. Rev. **D 71**, 112005 (2005), and references therein.
11. E. Aliu et al, K2K Coll. Phys. Rev. Lett. **94**, 081802 (2005).
12. "Neutrino Astrophysics", J.N. Bahcall, Cambridge University Press (1989).
13. Q.R. Ahmad et al. SNO Coll., Phys. Rev. Lett. **89** 011301 (2002); B. Aharmim et al. SNO Coll., Phys. Rev. **C 72**, 055502 (2005).
14. J.N. Bahcall, A.M. Serenelli and S. Basu, Astrophys. J. **621**, L85 (2005), and references therein.
15. K. Eguchi et al., KamLAND Coll., Phys. Rev. Lett. **90**, 021802 (2003). T. Araki et al., KamLAND Coll., Phys. Rev. Lett. **94**, 081801 (2005).
16. A. Aguilar et al., LSND Coll., Phys. Rev. **D 64**, 112007 (2001).
17. M. Apollonio et al., CHOOZ Coll., Phys. Lett. **B 466**, 415 (1999).
18. B.D. Fields and S. Sarkar, in Ref. [1], J. Phys. **G 33**, 220 (2006), and references therein.

19. R.H. Cyburt et al., Astropart. Phys. **23**, 313 (2005).
20. U. Seljak, A. Slosar and P. McDonald, JCAP **0610**, 014 (2006).
21. D.G. Michael et al., MINOS Coll., Phys. Rev. Lett. **97**, 191801 (2006).
22. H.T. Wong, Mod. Phys. Lett. A 20, 1103 (2005), and references therein.
23. H.T. Wong et al., TEXONO Coll., Phys. Rev. **D 75**, 012001 (2007).
24. B. Xin et al., TEXONO Coll., Phys. Rev. **D 72**, 012006 (2005).
25. H.M. Chang et al., TEXONO Coll., Phys. Rev. **D 75**, 052004 (2007).
26. T. Araki et al., KamLAND Coll., Nature **436**, 499 (2005).
27. R. Nahnhauer, in Ref. [3], Nucl. Phys. **B** (Procs. Suppl.) **143**, 387 (2005), and references therein.
28. H.T. Wong et al., J. Phys. Conf. Ser. **39**, 266 (2006)

SPRINGER PROCEEDINGS IN PHYSICS

81 **Materials and Measurements in Molecular Electronics**
Editors: K. Kajimura and S. Kuroda

82 **Computer Simulation Studies in Condensed-Matter Physics IX**
Editors: D.P. Landau, K.K. Mon, and H.-B. Schüttler

83 **Computer Simulation Studies in Condensed-Matter Physics X**
Editors: D.P. Landau, K.K. Mon, and H.-B. Schüttler

84 **Computer Simulation Studies in Condensed-Matter Physics XI**
Editors: D.P. Landau and H.-B. Schüttler

85 **Computer Simulation Studies in Condensed-Matter Physics XII**
Editors: D.P. Landau, S.P. Lewis, and H.-B. Schüttler

86 **Computer Simulation Studies in Condensed-Matter Physics XIII**
Editors: D.P. Landau, S.P. Lewis, and H.-B. Schüttler

87 **Proceedings of the 25th International Conference on the Physics of Semiconductors**
Editors: N. Miura and T. Ando

88 **Starburst Galaxies**
Near and Far
Editors: L. Tacconi and D. Lutz

89 **Computer Simulation Studies in Condensed-Matter Physics XIV**
Editors: D.P. Landau, S.P. Lewis, and H.-B. Schüttler

90 **Computer Simulation Studies in Condensed-Matter Physics XV**
Editors: D.P. Landau, S.P. Lewis, and H.-B. Schüttler

91 **The Dense Interstellar Medium in Galaxies**
Editors: S. Pfalzner, C. Kramer, C. Straubmeier, and A. Heithausen

92 **Beyond the Standard Model 2003**
Editor: H.V. Klapdor-Kleingrothaus

93 **ISSMGE**
Experimental Studies
Editor: T. Schanz

94 **ISSMGE**
Numerical and Theoretical Approaches
Editor: T. Schanz

95 **Computer Simulation Studies in Condensed-Matter Physics XVI**
Editors: D.P. Landau, S.P. Lewis, and H.-B. Schüttler

96 **Electromagnetics in a Complex World**
Editors: I.M. Pinto, V. Galdi, and L.B. Felsen

97 **Fields, Networks, Computational Methods and Systems in Modern Electrodynamics**
A Tribute to Leopold B. Felsen
Editors: P. Russer and M. Mongiardo

98 **Particle Physics and the Universe**
Proceedings of the 9th Adriatic Meeting, Sept. 2003, Dubrovnik
Editors: J. Trampetić and J. Wess

99 **Cosmic Explosions**
On the 10th Anniversary of SN1993J (IAU Colloquium 192)
Editors: J. M. Marcaide and K. W. Weiler

100 **Lasers in the Conservation of Artworks**
LACONA V Proceedings, Osnabrück, Germany, Sept. 15–18, 2003
Editors: K. Dickmann, C. Fotakis, and J.F. Asmus

101 **Progress in Turbulence**
Editors: J. Peinke, A. Kittel, S. Barth, and M. Oberlack

102 **Adaptive Optics for Industry and Medicine**
Proceedings of the 4th International Workshop
Editor: U. Wittrock

103 **Computer Simulation Studies in Condensed-Matter Physics XVII**
Editors: D.P. Landau, S.P. Lewis, and H.-B. Schüttler

104 **Complex Computing-Networks**
Brain-like and Wave-oriented Electrodynamic Algorithms
Editors: I.C. Göknar and L. Sevgi

105 **Computer Simulation Studies in Condensed-Matter Physics XVIII**
Editors: D.P. Landau, S.P. Lewis, and H.-B. Schüttler

MIX
Papier aus verantwortungsvollen Quellen
Paper from responsible sources
FSC® C105338

If you have any concerns about our products,
you can contact us on
ProductSafety@springernature.com

In case Publisher is established outside the EU,
the EU authorized representative is:
Springer Nature Customer Service Center GmbH
Europaplatz 3, 69115 Heidelberg, Germany

Printed by Libri Plureos GmbH
in Hamburg, Germany